Sculptors of the Universe

Life's Creative Role in a Chaotic Cosmos

By Arnon Daniel Katz

An exploration of life's ability to carve patterns, create structure, and transform the universe, one fleeting moment of creativity at a time.

"Only entropy comes easy"

Anton Chekhov

Copyright © 2024 by Arnon Daniel Katz

All rights reserved. No part of this book may be reproduced, stored in a retrieval system, or transmitted in any form or by any means—electronic, mechanical, photocopying, recording, or otherwise—without prior written permission from the publisher, except for brief quotations in critical reviews or articles.

Edition: 2024

ISBN: 979-8-30425-812-8

Imprint: Independently published

Note:

This book is not intended to serve as a scientific document. The ideas, concepts, and interpretations presented are solely those of the author and reflect personal insights and perspectives. Readers are encouraged to explore further and draw their own conclusions.

To enhance your reading experience, I've added a playlist of related videos on YouTube that further explore the concepts in this book—featuring expert interviews, guided exercises, and more. Feel free to scan the QR codes to watch them and deepen your understanding.

Explore Further on YouTube

As these topics continue to evolve, I'll update the book—So please **follow me on Amazon** to stay tuned.

Last thing, Please feel free to share your thoughts with me via my **Telegram**. Use this QR code to connect with me:

Thank you for joining me. I hope it sparks your curiosity and inspires you to keep exploring!

Table of Contents
Introduction: Life's Whisper
Part I: The Framework of Creation
Chapter 1: Entropy—The Universe's Tendency Toward Chaos
Chapter 2: Time's Arrow—Entropy and the Evolution of the Cosmos
Part II: Life's Sculpting Hand
Chapter 3: Seeds of Structure—Life's Defiance of Disorder
Chapter 4: Quantum Creativity—The Genius of Green Leaves
Part III: The Shared Symphony of Life
Chapter 5: Humanity's Canvas—Sculptors of the Earth and Beyond
Chapter 6: Cosmic Sculptors—Life Beyond Earth
Chapter 7: Universal Principles—Myths, Science, and Meaning
Part IV: Life's Endless Possibilities
Chapter 8: Complexity Unleashed—Evolution as a Creative Force
Chapter 9: The Sculptor's Invitation—A Creative Future
Conclusion
Appendices: Tools for Understanding
Bibliography: Diving Deeper
Inspirational Figures
Conclusion of the Appendices

Introduction: Life's Whisper
The universe, at a glance, seems like a place driven mostly by blind, indifferent forces. Stars ignite and explode without regard for who's watching; galaxies drift through the darkness, scattering their cosmic seeds haphazardly. In this grand stage, governed by the subtle tug-of-war between order and disorder, it might appear that life is just an afterthought—an improbable flicker hidden in the cosmic void. Yet, if we look closely enough, we discover that life is not merely a passive bystander. Life whispers to the cosmos, gently suggesting new forms of complexity, quietly sculpting fresh layers of order from the raw clay of chaos.

In a world ruled by entropy—the steady decay of structure into randomness—life stands as a remarkable counterforce. Like a potter's wheel spinning clay into elegant shapes, living systems generate complexity from an unyielding torrent of disorder. Their strategies vary, ranging from the microscopic genius of green leaves capturing photons to the grand orchestration of entire ecosystems balancing energy and matter in delicate harmony. Life, in all its forms, continuously rearranges the cosmic puzzle.

But it is humans, those curious and ever-inventive creatures, who take this sculpting impulse to unprecedented heights. We have charted deep oceans, launched telescopes beyond Earth's boundaries, and crafted technologies that allow us to manipulate matter and energy as no other species can. With these tools, we reshape landscapes, design intricate machines, and even edit the code of life itself. Our presence is a testament to something astonishing: the universe has not only given rise to observers—it has birthed creators.

This book is a journey into the heart of that creative dance. We will weave together threads from physics, biology, philosophy, and psychology to paint a tapestry of understanding. We'll explore how entropy sets the stage, how time's unrelenting arrow propels the cosmos forward, and how life emerges as a sculptor, a defier of chaos. We'll watch as tiny seeds push through cracks in asphalt, transforming barren cityscapes into green oases, and we'll dive into the hidden realm of quantum mechanics, where chlorophyll molecules perform calculations that rival any human engineer.

In these pages, we'll also question what it means for humanity to be both a child of the stars and a shaper of worlds. We'll consider what happens if we're not alone in the universe—if life elsewhere also sculpts complexity amidst chaos. Along the way, we'll uncover how myths and scientific theories echo each other, how ancient stories and modern data converge on the same profound truth: life is not just a byproduct of the cosmos, but an active participant in its unfolding.

So, take a moment to listen closely. The whispers of life are all around us. Let's tune our ears and open our minds to the exquisite interplay of creation and disorder, complexity and simplicity, innovation and inevitability.

As we continue our journey, let's consider the delicate balance that life must strike to persevere in a universe seemingly biased toward decay. On a cosmic scale, stars burn their fuel and collapse, planets bear the scars of impacts, and even the grand architecture of galaxies slowly unravels. Yet, within these islands of order—where matter and energy find fleeting alignments—life nurtures patterns that challenge the dominance of disorder. It's as if life has found a secret toolkit, enabling it to build castles of complexity in the sand, even as the cosmic tide threatens to wash them away.

This doesn't mean life somehow "defeats" entropy in an absolute sense. The laws of thermodynamics remain unbroken, and the grand cosmic river still flows downhill. Yet life has learned to channel tiny eddies and currents within that river, creating temporary whirlpools where order can take shape. Consider a small seed: inside its hardened shell, genetic instructions lie dormant, waiting for the right conditions—moisture, warmth, a crack of sunlight. When these conditions are met, the seed springs into action, orchestrating the construction of roots, leaves, and stems. It's an exquisite dance of matter and energy made possible by the subtle manipulations life has honed over billions of years.

If we look closer, life's art lies in its dynamic ability to find new pathways in a universe that often drifts toward dissolution. On the grand scale, stars exhaust their fuel, planets bear the scars of ancient collisions, and galaxies slowly shift in ways that can spread matter thin. And yet, between these unfolding dramas, life continues to thread delicate patterns, forging intricate structures where disarray might otherwise prevail. Rather than surrendering to disorder, life probes the boundaries of possibility, adapting and innovating in ways that defy easy predictions.

We might imagine life's role as that of a sculptor working with a restless medium—an endlessly changing block of clay that dries unevenly, cracks unpredictably, and rarely holds a shape for long. Yet the sculptor need not resign itself to watching each masterpiece crumble. With each new attempt, techniques improve. The sculptor refines methods, discovers sturdier materials, and learns how to preserve form in the face of shifting conditions. Over countless experiments, life's sculpting hand may find ways to keep complexity standing taller, lasting longer, and evolving ever more gracefully.

It's not certain that entropy must always have the final word. The known laws of physics haven't changed, and certainly the broad currents of cosmic change continue flowing. But within those currents, life has displayed an uncanny knack for seizing opportunities. It finds pockets of energy and reorganizes them into forests, reefs, tundras, and rainforests; into civilizations brimming with ideas, technologies, and grand ambitions. This is not a fixed dance where one partner always leads. Instead, it's a fluid waltz: entropy nudges toward dissolution, while life responds with unexpected leaps in complexity, forever testing the boundaries of what can endure.

Humanity's role in this unfolding drama brings a fresh twist to the narrative. Over millennia, we learned to harness fire and carve tools from stone. We cultivated fields that tamed wild plants, engineered marvels of architecture and infrastructure, and tapped into the hidden realms of biology and quantum mechanics. In doing so, we've become extraordinary co-creators with nature, wielding capacities our ancestors could scarcely imagine. Could we, guided by knowledge, creativity, and care, extend life's repertoire even further—finding new ways to stabilize complexity on greater scales, and for longer spans of time?

These questions become especially pressing when we contemplate the distant future. Is it possible that life, through continual innovation and cooperation, might someday set down roots that endure the shifting sands of cosmic time? Could the strategies life refines on Earth one day influence the fate of worlds far beyond our star, enabling an ever-richer tapestry of living systems to flourish?

By framing these questions, we invite reflection rather than resignation. The direction of this dance is not predetermined. As readers, thinkers, and participants in life's ongoing story, we're left to wonder: if there is no known limit to the inventiveness and resilience of living beings, can this ancient interplay tilt the balance? Could life, given enough patience and ingenuity, permanently shift the world's trajectory toward complexity, meaning, and lasting creation?

As we move forward, let's shift our focus to how we perceive this grand interplay between life and the cosmos. The idea that life might one day refine its craft to the point of redefining cosmic fate is bold and thrilling, but it's not without its doubts and dilemmas. Just as an artist never truly knows what a masterpiece might look like until it takes shape, we cannot be sure where the creative instincts of living beings will ultimately lead. Yet, there's something profoundly human in this uncertainty—in daring to hope that complexity can keep unfolding, that meaning can keep deepening, and that existence might bloom in ever more intricate patterns.

One key to understanding these possibilities lies in recognizing that life's creative strategies aren't limited to biological mechanisms alone. Culture, technology, imagination, and philosophy all play vital roles in how we extend life's influence. Our inventions, from simple tools to complex computational networks, serve as bridges, linking human ingenuity to larger natural patterns. They allow us to orchestrate matter and energy with increasing subtlety, building layered systems of organization that echo—and sometimes surpass—those found in nature.

Indeed, we've already begun to integrate life's sculpting tendencies with our own dreams. Our cities might be chaotic at times, but they're also teeming with complexity and interconnection. Our artistic traditions, scientific explorations, and ethical debates all contribute to the ongoing narrative of how life interacts with the universe. In these endeavors, we are effectively "sculpting" new niches within the cosmic tapestry. We are experimenting not just with physical forms, but with conceptual spaces, value systems, and visions of the future.

This isn't to say that humanity, or any other form of life, is guaranteed a triumphant destiny. The universe remains vast and mysterious, filled with forces we only partially understand. Disaster can strike from unexpected quarters—asteroid impacts, supernovae, or more subtle cosmic shifts that challenge life's hard-won footholds. On a more immediate level, our own shortsighted choices could disrupt the delicate balance we rely on, ushering in new waves of disorder. Yet, these uncertainties also provide fertile ground for growth. They remind us that the interplay of life and entropy is not a simple equation with a known solution, but rather an evolving dialogue—one in which every generation, every new idea, and every adaptation adds another stanza.

When we stand beneath a star-studded sky, we may feel small, but we are not powerless. Our ability to learn and adapt, to merge our minds with the wisdom embedded in nature, gives us a unique vantage point. We participate in the cosmic dance, contributing our steps, our gestures, our dreams. Perhaps the ultimate question is not whether life will "win" against entropy, but how far this dance can proceed and how many new forms it can inspire.

As we embark on this journey through the chapters ahead, we'll examine entropy's relentless pull, the subtle strategies life deploys to resist and reshape that pull, and the roles we humans play as both products and active participants in this ongoing story. Together, we'll explore how life's persistent whisper might one day become a resonant chorus, steering the cosmos toward vibrant harmonies that no single mind could have imagined. There may be no definitive final note—but that, in itself, may be the greatest promise of all.

Part I: The Framework of Creation

Chapter 1: Entropy—The Universe's Tendency Toward Chaos

Imagine a carefully arranged stack of dominoes, each one poised to tip the next in a delicate chain reaction. When they're all standing tall and perfectly aligned, the scene feels brimming with potential. But give the first domino a little push, and within moments, you're left with a scattered jumble on the floor. This shift from tidy order to messy chaos is something we've all seen, whether in toppled dominoes, a once-tidy desk gone askew, or a child's bedroom floor flooded with toys. Beneath the surface of these everyday images lies a grand principle that shapes the fate of everything from stars to teacups: entropy.

In a strict physical sense, entropy is often described as a measure of disorder. It's a concept that cuts across disciplines, surfacing in thermodynamics, information theory, and even cultural metaphors. At its core, entropy sets the stage for much of what we experience in the universe—its relentless drive toward greater randomness and less structure. Just as a dropped glass shatters into fragments rather than reassembling itself, the natural trend is for things to fall apart. Given time, even the most orderly systems succumb to entropy's steady pressure.

But it's crucial to realize that entropy isn't an evil force lurking in the shadows—it's simply the statistical tendency for systems to explore more probable states. In other words, it's much easier to be disorganized than it is to be neatly arranged. For every elegant arrangement of molecules or objects, there are vastly more ways for them to be jumbled. As a result, the universe doesn't really "prefer" chaos; it's just that chaos is the most likely outcome. If you gently place a drop of food coloring in a still glass of water, it diffuses outward until the water is evenly tinted. The dye doesn't cluster back together in one corner unless you intervene. This one-way journey—from order to disorder—is at the heart of entropy's story.

Entropy also plays a profound role in how we perceive time's arrow. We never see a shattered glass spontaneously pulling its shards back together, nor do we encounter scrambled eggs un-scrambling themselves. Such processes, while not impossible according to the fundamental laws of physics, are so staggeringly improbable that we consider them effectively forbidden. This one-directional flow—from less disordered to more disordered—helps give time its familiar shape. Yesterday's ice cubes, once neatly frozen, are now melted puddles. Tomorrow's state of affairs, whatever it holds, will likely be slightly more disordered than today's, adding another subtle chapter in the saga of entropy's rise.

Of course, the reality is not so simple. Entropy's steady increase doesn't mean complexity can't appear—life, as we've seen, can carve out pockets of order. Yet these pockets exist against an ever-rising tide, making their achievements all the more impressive. To understand the significance of this dynamic, we must dive deeper into what entropy truly means, how it operates, and how it underlies some of the universe's most fascinating phenomena.

To grasp entropy's power, let's turn our attention to a simple experiment you might perform in your kitchen. Take a hot cup of coffee and place it on a counter. Wait an hour and check its temperature. Unless you've got a peculiar heat source lurking nearby, you'll find that your once-steaming cup has cooled down significantly. But notice that it never goes the other way: a lukewarm coffee does not spontaneously reheat. This daily, unremarkable observation is another expression of entropy at work. Heat, which can be understood as a measure of energy spread, naturally flows from hotter objects to cooler surroundings, not the other way around. In fact, this simple rule of heat flow forms the backbone of the Second Law of Thermodynamics, often described in layman's terms as "entropy increases over time." Whenever energy disperses from a concentrated form into a more widely distributed one, entropy steps in to guide the process. From this perspective, the universe itself can be seen as an enormous stage where energy, once more orderly and concentrated shortly after the Big Bang, has gradually spread out, diluting into the vastness of space. The birth of stars and galaxies took shape in the early chapters of the cosmos, but as eons pass, the cosmos is expected to move toward an ever more diffused and less structured state.

Yet, it's important to emphasize that entropy is not synonymous with chaos in some vague, destructive sense. Rather, it's a statistical reality. Think about rearranging your living room furniture. There are countless ways the furniture can be scattered about, and only a few arrangements that feel "just right." The likelihood of randomly stumbling into a neat configuration is slim. By analogy, the universe has vastly more "messy" states than neat ones. Without any guiding hand continually working to maintain order, the cosmic furniture naturally ends up, so to speak, all over the place.

This is where life's defiance of entropy becomes truly riveting. Life doesn't break the rules—no living organism can create a perpetual motion machine or reverse the natural flow of energy. But what life can do is tap into external energy sources and direct them into building and maintaining complex structures. Take the green leaf, for instance. By capturing sunlight and using it to power photosynthesis, the leaf keeps its internal machinery more ordered and well-regulated than if it passively submitted to the universe's trend toward disorganization. It's a localized balancing act, a temporary pushback against the cosmic tide of entropy.

In this interplay, entropy sets the boundaries, and life tries to find loopholes, or at least small sheltered harbors of complexity. By understanding entropy, we gain insights into why perpetual motion machines are impossible, why time seems to move forward, and why life's remarkable feats of creativity stand out against a backdrop that continually nudges everything toward random scattering. As we continue our exploration, we'll see that entropy is neither villain nor ally. It's simply the stage on which our cosmic drama unfolds—an influence we must understand to appreciate the clever dance that life and the universe perform together.

As we widen our perspective, entropy's role grows ever more profound. Cosmologists often describe a far-future scenario for the universe, one in which stars have burned out, black holes have evaporated, and matter is spread thinly across an almost featureless expanse. In such a universe, there would be no bright galaxies, no towering nebulae, no swirling storms of gas—just an even haze where everything is lukewarm and unchanging. This hypothetical "heat death" of the universe represents the ultimate triumph of entropy, with all energy dispersed so thoroughly that nothing interesting can happen. The cosmos would be a place devoid of gradients or contrasts, offering no opportunities for complexity or life.

But that grim picture, while consistent with our current understanding of physics, is far from the whole story. We are living now, in a universe that's still relatively young, filled with dazzling phenomena, from supernovae to humming rainforests. The entropy-driven journey toward uniformity isn't happening in a straight line. Instead, it's playing out in dynamic pockets of resistance and innovation. Life on Earth has managed to carve out niches that are anything but static—verdant jungles, bustling coral reefs, and evolving cultures that generate music, art, and sophisticated science. All these blossoms of complexity are supported by continuous flows of energy, like sunlight fueling photosynthesis or geothermal vents sustaining deep-sea ecosystems.

On a human scale, understanding entropy teaches us humility. It helps us see that maintaining order—whether in our personal lives or our societies—requires constant effort. A clean room does not stay clean without regular tidying, just as a functioning civilization cannot thrive without the maintenance of infrastructure and the nurturing of shared values. Order is something that must be earned and renewed; it does not remain stable on its own. In this sense, entropy invites us to appreciate the fragile beauty of the structures we rely upon.

Yet, entropy doesn't just limit what's possible; it also catalyzes creativity. Because nature tends toward mixed-up states, the emergence of order feels almost like a rebellion—an unexpected feat that draws our admiration and curiosity. The improbable alignment of molecules into a living cell, or the carefully orchestrated dance of neurons forming a human thought, stands out against entropy's backdrop, revealing that complexity can thrive when given the right circumstances. By understanding entropy, we become more aware of the delicate conditions required to sustain complexity and more inspired to preserve and enhance those conditions.

As we venture into subsequent chapters, we'll explore how time itself is shaped by entropy's relentless push, how life devises clever strategies to resist disorder, and how humans extend these strategies into realms of technology and culture. The concept of entropy frames the cosmic puzzle—its rules, constraints, and eventual outcomes. But life, in its myriad forms, represents a counterpoint—a sculptor's hand molding new shapes, a rebellious current swimming against the tide.

In embracing this tension, we begin to see not just a universe running down but also a tapestry where bright threads of complexity weave vibrant patterns into the grand design. Our story is not simply one of decline but of dynamic interplay, where life's whispers might someday grow louder, more confident, and more capable of redefining what's possible.

Chapter 2: Time's Arrow—Entropy and the Evolution of the Cosmos

Imagine standing on a winding road with no signs, no markers—just the wide-open horizon. Without guidance, you might wander back and forth without any sense of direction. But in our universe, there's an invisible compass that points ever-forward: time's arrow. This arrow is not etched in stone, nor is it pulled taut by some cosmic archer. Instead, it emerges naturally from the universe's laws. And at its core, entropy—the measure of disorder—helps set the direction.

We experience time as a one-way street. We recall the past, not the future; we know what we had for dinner last night, but not what we'll eat tomorrow. Yet, the fundamental equations of physics, when viewed in isolation, don't demand time to flow in one direction or another. On paper, these laws work just as well backward as forward, at least in theory. So why do we witness a reality where eggs crack but never spontaneously reassemble, where memories accumulate in one temporal direction, and where our coffee grows cold rather than hot?

The answer lies in the statistical nature of entropy. Early in the universe's history—a fraction of a second after the Big Bang—matter and energy were arranged in a remarkably low-entropy state. This means, in some sense, the cosmos began its life in a comparatively neat and orderly arrangement. As the universe expanded, it had countless more "messy" configurations available than neat ones. Over cosmic time, matter and energy naturally drifted into those messier states, just as a tidy room tends to become cluttered. This grand drift sets a temporal gradient: the past is remembered as more orderly, while the future promises increasingly scattered arrangements. Time, in essence, is the unfolding story of entropy's steady rise.

From stars burning their hydrogen fuel to black holes stirring the cosmic stew, every event nudges the universe toward states of greater entropy. The swirling dance of galaxies, the birth of complex molecules, the formation and dissolution of life's fragile patterns—all are threaded onto this arrow of time, carried forward by the current of increasing disorder. Without this gentle push, time could, in theory, run either way, leaving no meaningful distinction between yesterday and tomorrow.

But let's be careful not to reduce time's arrow to a bleak narrative of unavoidable breakdown. Rising entropy may define the direction of time, but along that path, complexity and creativity can flourish. In fact, it's precisely because the universe started off in an exceptionally ordered state that we've been able to witness the blossoming of stars, galaxies, planets, and life. The story of time's arrow is not just one of decay; it's also the tale of emergent structures and surprising inventions—like a symphony whose subtle shifts in melody rely on an underlying rhythm.

This interplay between entropy and time's arrow grants our world its storytelling nature. Each moment can be understood as a turning of the cosmic page, a chapter in the drama of creation and dissolution, growth and letting go.

As we dig deeper, we find that time's arrow shapes not only the grand, cosmic narrative but also the intimate details of our daily lives. Consider why we remember the past but not the future. Memory itself is a record etched into complex neural networks—an arrangement of molecules and synapses that demands energy and structure. For that structure to form, information must be imprinted in a specific way that reduces the uncertainty about past states of the world. This imprinting process happens as we move from lower entropy states to higher ones. In other words, it's feasible to store traces of where we've been, but not where we haven't yet gone. The rules favor past information sticking around because we were once in a simpler, more constrained state, from which these memories can emerge as the universe unfolds.

Outside our skulls, time's direction reveals itself in subtle yet pervasive ways. Footprints on a sandy beach tell the story of someone who was there. The waves may wash them away, increasing the disorder of disturbed sand grains and erasing this neat arrangement—but they won't spontaneously re-form the pattern of a human foot. The trail of fossils buried in sedimentary rock layers, the weathered faces of mountains, the peeling paint on an old fence—these are all records that point one way. Our entire idea of history, of a narrative unfolding, is anchored in the fact that we can trace lines backward through an increasingly ordered record and forward into a future of greater uncertainty.

Moreover, time's arrow influences how we think about causality. We intuitively understand that causes precede effects—push a book off a table, and it will fall, not the other way around. This causal ordering dovetails with entropy's nudge toward disorder. Events that scatter energy and matter—like the falling book—generate more likely subsequent states than events that would gather scattered pieces back together. This alignment between causality and entropy growth isn't coincidental; it's woven into the very fabric of how we perceive reality and make sense of cause and effect.

Yet, for all this talk about inevitability, we should not overlook life's capacity to innovate within the rules. The arrow of time sets a broad direction, but life and human creativity find clever ways to eke out complexity and meaning from the journey. Life exploits energy flows—like those from the Sun—to build improbable structures and maintain intricate patterns. Our technologies and ideas do something similar, arranging knowledge and matter into new configurations that seem unlikely amid the cosmic drift toward chaos.

In doing so, we humans have begun to wonder: Could time's arrow ever bend or shift? What if, by harnessing the laws of nature more profoundly, we discovered strategies to stabilize complexity over longer periods, or even direct the flow of events to create unlikely outcomes? While these remain distant conjectures, they remind us that we are active participants in this narrative. Our understanding of time's arrow may not just be observational; it might someday become a tool we wield to shape the future.

At the largest scales, time's arrow isn't simply a curious quirk of human perception; it's a defining characteristic of how the universe evolves. In the cosmic tapestry, structures form as gravity clumps matter together, forging stars that convert light elements into heavier ones. Galaxies swirl in elaborate spirals, while planetary systems arise from swirling disks of dust and gas. Each act of formation takes energy and organizes it in complex ways. Yet the overarching trend, as stars burn their fuel and release energy into space, is toward a more even dispersal of matter and radiation. The arrow of time ensures that, although beauty and complexity can flourish, they do so against a backdrop where overall order inevitably slips away.

For all its constraints, however, time's arrow also grants a kind of narrative coherence to the universe. Without it, existence would feel like a film reel that could be played forward or backward with equal meaning. Instead, we get storylines: from the Big Bang to the birth of galaxies, from the formation of planets to the emergence of life. Complexity arises as a byproduct of these transitions, allowed to bloom like vibrant flowers even if, in the grand tally, the garden's soil becomes ever more scattered with each passing season.

In a sense, time's arrow creates the conditions necessary for life to perceive, react, and ultimately shape its environment. Organisms develop senses and memories because, in a universe where entropy increases, there's value in knowing where things have been and predicting where they might go. Our capacity for foresight emerges precisely because the future is less constrained than the past. We can imagine building structures, alliances, and knowledge systems that anticipate entropy's relentless push, devising ingenious ways to surf the wave of disorder rather than being overwhelmed by it.

This perspective hints at deeper philosophical and psychological questions. How much of our cultural understanding—our sense of purpose, morality, and creativity—arises from the existence of time's arrow? Would meaning feel so poignant if events could just as easily run backward, erasing the narratives we cherish? By increasing entropy, the universe nudges us to witness things break down, move forward, and transform. Within that cycle, life's persistent attempts at self-organization and creation stand out as remarkable counterpoints.

As we move through the chapters to come, we'll continue exploring how entropy and time's arrow shape our world and frame life's possibilities. We'll delve into how life not only emerges despite rising disorder but thrives by playing off it, discovering strategies to sustain complexity. We'll look at how humans, equipped with tools, technology, and imagination, might extend life's hold and even influence the future path of cosmic evolution. Ultimately, time's arrow isn't just a one-way street—it's the track upon which the universe's greatest stories unfold, offering us the chance to become co-authors as we learn its language and unlock its secrets.

Part II: Life's Sculpting Hand

Chapter 3: Seeds of Structure—Life's Defiance of Disorder

Picture a tiny seed nestled in the crevice of an old stone wall. This seed, scarcely bigger than a pinhead, holds the blueprint for an entire organism—roots, stems, leaves, and maybe even flowers that will one day burst forth into the world. At first glance, this seems remarkable. How can so much order emerge from something so small and simple, especially in a universe that seems to prefer entropy and chaos at every turn?

This miniature miracle offers a clue about life's profound strategies for carving out structure. Even as entropy nudges the cosmos toward disorder, living beings persistently create intricate forms, building complexity out of raw material and energy. Seeds are just one example of life's ongoing attempt to assert its own patterns within the world's broader turbulence. They are nature's time capsules, carrying genetic information across seasons and generations, poised to initiate a new chapter of growth whenever conditions prove favorable.

The challenge of maintaining order against the pull of entropy is immense. The molecules inside a seed do not spontaneously organize themselves into a functional plant. Instead, life relies on stored instructions DNA and a carefully orchestrated sequence of chemical reactions. By tapping into external energy sources, like sunlight or the nutrients in soil, life essentially "tricks" the universe into allowing tiny enclaves of order to form and endure. Each seed's germination, each sprout breaking through the soil's crust, is a testament to the quiet resilience and ingenuity of life's methods.

It's essential to note that no violation of physical laws occurs here. Life's defiance is subtle and compliant with the grand rules of thermodynamics. Indeed, the order within an organism comes at a cost: life must constantly import energy and export entropy into its surroundings. In other words, the plant that grows from a seed helps spread disorder elsewhere—perhaps by breaking down nutrients from the soil or releasing heat—while carefully preserving and enhancing order within its own tissues. The end result is a delicately balanced dance, one that allows local complexity to increase even as the universe's overall entropy marches onward.

Seeds of structure can be found everywhere in the living world. Single-celled microbes, coral reefs, towering sequoias, and intricately designed spiderwebs all represent life's knack for turning chaos into complexity. What ties these diverse examples together is a common theme: life finds ways to channel energy efficiently, using it to stabilize and expand its architecture. In doing so, living organisms become architects of order, sculpting improbable forms from the raw clay of nature's building blocks.

Yet, the struggle doesn't end once the seed germinates and the plant grows tall. Life's defiance of disorder isn't a final victory—it's an ongoing negotiation, a never-ending series of clever adaptations. In the chapters ahead, we'll see just how creative life's solutions can become, especially when we turn our gaze toward the hidden world of quantum mechanics, photosynthesis, and human ingenuity itself.

Step back for a moment and consider how life's methods developed. The earliest organisms to emerge on Earth had none of the grandeur we see in towering forests or bustling coral reefs. They were microscopic pioneers, crude by modern standards, yet they carried within them the spark of complexity. Over eons, evolution refined the genetic codes, nudged chemical networks toward greater sophistication, and allowed for increasingly clever ways to harness energy. Each adaptation, each innovative twist, allowed life to sculpt new patterns and survive in environments that seemed inhospitable.

From these humble beginnings, every ecosystem we know today can trace its lineage. A seed, after all, is just a vessel for genetic knowledge gathered by its ancestors over millions of years. Plant lineages that successfully navigated drought, floods, parasites, and predators passed along finely tuned instructions. Over time, these instructions expanded into a rich dictionary of solutions—biochemical shortcuts, defensive tactics, and metabolic tricks—that give new generations a head start in their battle against disorder.

Consider how seeds remain dormant until just the right conditions emerge. This pause button on life's complexity illustrates another hallmark of life's defiance: patience and timing. The seed doesn't waste its precious stores of energy trying to sprout during a drought or a deep winter freeze. Instead, it waits, quietly holding order within its shell. Once the environment cooperates, the seed taps into moisture, warmth, and nutrients, launching a grand architectural project on a cellular scale—cells dividing, tissues differentiating, leaves unfurling. It's a silent symphony, guided not by conscious intent but by evolutionary memory encoded in each cell's DNA.

Of course, seeds are not alone in this game. Around the world, different life forms have discovered their own methods to tip the scales in their favor. Some species store energy as fat, waiting out lean times. Others rely on symbiotic relationships—fungi partnering with plant roots, bacteria teaming up within animal digestive tracts—to share resources and maintain order. These partnerships can enhance resilience, allowing organisms to stand stronger against entropy's pull.

Humanity's influence in this narrative has grown ever more pronounced. Through agriculture, we have harnessed seeds to feed entire civilizations. By selecting which plants to cultivate and which traits to encourage, we have become co-creators in the story of life's defiance. Our efforts have helped certain species flourish far beyond their natural ranges. Fields of wheat, corn, and rice are testaments to our role as active participants in the dance between structure and disorder. Yet, our interventions come with responsibilities and risks, as the systems we create must still reckon with entropy's demands, and the complexity we foster can be fragile.

In a universe that seems destined to drift toward chaos, life's ability to push back is a gentle reminder that not all outcomes are written in stone. Seeds, along with every other living thing, represent life's stubborn refusal to surrender to entropy. They show us that complexity can emerge, endure, and evolve—drawing strength from cooperation, adaptation, and the tireless search for new ways to wring order from chaos.

As we step further into the subtle art of life's defiance, it's worth reflecting on how energy flows underwrite every success story. Seeds rely on careful metabolic accounting, turning light, water, and soil nutrients into organized tissues. Elsewhere, microbial communities collaborate to break down complex substances, weaving them into more useful forms. Each of these examples demonstrates how life exploits energy differentials—using them as stepping stones to ascend against entropy's slope.

Evolution, too, can be seen as a kind of energy-driven learning process. Mutations that improve an organism's ability to harness energy and maintain structure are favored over time. This selective pressure fine-tunes life's "playbook," generating a diverse and adaptable arsenal of strategies. Just as a musician refines their technique over years of practice, life refines its approach to resisting disorder, making slight improvements generation by generation.

It's important, however, not to romanticize the struggle too much. Life's defiance is a constant work-in-progress, and entropy never truly goes away. Organisms must continually expend energy to keep their internal systems from unraveling. Cells repair damaged DNA, replace worn-out proteins, and defend themselves against invasive microbes. On a grander scale, ecosystems shuffle species in and out, adapting to climate shifts and the introduction of new competitors. Life's triumphs are always conditional, contingent on the delicate balance between available energy and the relentless pull of disorder.

From a human perspective, understanding this tension can be both humbling and empowering. It's humbling because it reminds us that complexity is not given freely—it must be earned. Our bodies, minds, cultures, and societies all represent victories over entropy, maintained through constant input of work and resources. But this knowledge also empowers us by revealing that we belong to a lineage of problem-solvers. The same evolutionary heritage that endowed seeds with their ingenious timing and architecture has equipped us with the cognitive tools to innovate, build, and create.

In essence, seeds and other life-forms teach us that the future need not be bleak. Even as the universe tilts toward greater entropy, life insists that complexity is possible. This hopeful message resonates on a philosophical level. When we plant gardens, raise children, or invent new technologies, we're participating in the same cosmic dialogue that seeds have been engaged in for billions of years. We're finding ways to gather, transform, and preserve structure in a world inclined to scatter it.

Our own efforts—engineering resilient crops, restoring damaged habitats, or designing sustainable energy systems—further extend this legacy. By aligning our actions with the principles life has long employed, we may push back entropy's encroachment and foster complexity on even grander scales. In doing so, we transcend mere survival, moving toward a future where creativity, diversity, and meaning continue to blossom.

As we turn to the next chapters, we'll examine how life's strategies only become more ingenious at the quantum level, and how human ingenuity, informed by this understanding, can echo those same patterns of defiance. In the grand interplay between order and chaos, life's seeds of structure persist—guiding us toward ever more intricate possibilities.

Chapter 4: Quantum Creativity—The Genius of Green Leaves

Close your eyes and picture the gentle light of a morning sun filtering through the leaves of a towering tree. Those leaves, each a delicate membrane of life, are performing one of the most elegant and baffling feats in nature: they are turning sunlight into the building blocks of existence. Photosynthesis, that humble process taught in elementary school, is more than just a chemical reaction. Recent research suggests it rests, at least in part, on a quantum trick—a subatomic shortcut that helps plants and algae capture energy with astonishing efficiency.

At first glance, it might seem surprising that quantum mechanics, with its strange probabilities and particle-wave dualities, has any role to play in the everyday miracles of life. We usually think of quantum effects as confined to laboratories or exotic cosmic scenarios, far removed from anything as commonplace as a leaf. And yet, evidence has emerged that within the chloroplasts—those tiny green powerhouses inside plant cells—nature is harnessing quantum processes to optimize energy flow. This is not some esoteric flourish; it's a practical advantage that gives plants their remarkable ability to convert sunlight into chemical energy at near-perfect rates.

To appreciate this, consider how complex the energy capture process really is. When a photon of sunlight strikes a molecule of chlorophyll, that molecule becomes excited, temporarily boosted into a higher-energy state.

The trick is to guide this energy to a central hub—the reaction center—where it can be locked into chemical bonds and stored for later use. But the molecule that first catches the photon isn't always located right next to the reaction center; in fact, the energy must often hop across a network of pigment molecules to reach its final destination. If these hops were random, like a traveler wandering unfamiliar streets, a good portion of energy would be lost along the way.

Here's where quantum mechanics comes in: instead of stumbling through the network, the energy wavefunction spreads out and effectively explores all possible pathways simultaneously. By doing this, it can find the most efficient route, collapsing onto the best path before any significant loss occurs. This quantum walk through the molecular maze ensures that energy is transferred with remarkable precision, minimizing entropy's chance to swallow it up. In essence, nature has stumbled upon a solution that engineers and scientists have been striving to replicate—an optimal wiring diagram for energy flow.

Of course, plants aren't "aware" of quantum mechanics, nor do they consciously design their molecular systems. Rather, evolution has found an arrangement of pigments and proteins that exploits these subatomic rules. Just as seeds carry forward life's hard-earned lessons, leaves carry forward these subtle design principles, passing them on through millions of years of trial and error. The result is nothing short of extraordinary: a green quantum machine quietly at work in our gardens and forests, sustaining the chain of life.

To fully appreciate the marvel playing out in each leaf, consider the contrast between nature's solution and our own human-engineered systems. Solar panels, for all their sophistication, convert only a fraction of the sunlight that hits them into usable electricity. Much of the energy is lost as heat, scattered by imperfections or inefficiencies in the panels' design. Meanwhile, plants seem to nimbly dodge these pitfalls, channeling captured energy with an artistry that leaves our best technologies in the shade. The fact that quantum mechanics—a domain we consider strange and elusive—could be a key player here suggests we may still have much to learn from life's subtle engineering.

In a sense, the leaf is serving as a quiet mentor, demonstrating principles that might guide future innovations. Engineers are already looking to biomimicry—studying how nature solves problems at every scale—to inspire more efficient technologies. If we can understand the quantum mechanisms that allow leaves to transport energy so elegantly, we might design solar cells that mimic this process, dramatically increasing our ability to harvest renewable energy. Such advances wouldn't just be a nod to the genius of green leaves; they could help us tackle pressing global challenges like climate change and energy scarcity.

The presence of quantum effects in something as familiar as a leaf also nudges us to rethink the boundaries between life's daily activities and the deepest laws of physics. We often separate biology from physics, as if cells follow one set of rules and subatomic particles another. Yet, as we see here, the layers are intertwined. The plant leaf draws upon quantum weirdness—superposition, coherence, and other phenomena that rarely show up at macroscopic scales—to outsmart entropy's attempts to scatter precious energy. Nature, it seems, is not just a passive victim of cosmic rules but an ingenious navigator, finding loopholes and shortcuts in the laws that shape our reality.

This insight resonates with a theme that's been building throughout our exploration: life's strategies often involve clever manipulations of fundamental processes. Just as seeds leverage the storage of genetic information to launch new structures, leaves leverage quantum effects to manage energy flows. Both strategies underscore that complexity isn't just something that appears despite the laws of nature—it emerges by working intimately within those laws, discovering subtle tricks along the way.

For humans, understanding these tricks could mean far more than improving solar panels. It may reshape how we view our own pursuit of order and meaning. If a leaf can thread a quantum needle to sustain life's intricate web, perhaps we can learn to apply similarly inspired thinking to the challenges we face. Our attempts at complexity—be they cultural, technological, or ecological—might benefit from observing how nature maximizes possibilities and minimizes waste.

In the chapters ahead, we'll see how humanity, as both students and sculptors, has begun to experiment with these quantum lessons. But before we get there, it's worth pausing to acknowledge the elegance hidden in each leaf's green glow. Here, in the simplest of gestures—sunlight captured, energy guided with near-perfection—we find life's most subtle and profound workbench, a place where nature's creative genius unfolds beyond what we'd ever expect.

As we ponder the leaf's ingenious methods, a broader question emerges: how did such quantum-level finesse evolve? Evolution does not plan ahead or reason about efficiency. Instead, it operates through trial and error, sifting through countless variations over immense spans of time. Some early photosynthetic organisms may have stumbled into molecular arrangements that happened to allow quantum coherence—those peculiar states where particles act like waves and sample multiple pathways at once. These lucky variants would have captured energy more efficiently, leaving more offspring and ultimately dominating the gene pool.

Over generations, subtle tweaks to the arrangement of pigments, proteins, and reaction centers refined this quantum choreography, raising efficiency inch by inch. The result is a set of molecular structures that almost seem designed for their task, even though no conscious hand ever guided their formation. It's a triumph of nature's blind tinkering—an evolutionary masterpiece that harnesses the strangeness of subatomic physics to sustain life's grand networks.

This evolutionary narrative reminds us that even the most baffling tricks need not be the product of foresight. Life's strategies often arise from mere happenstance. What's remarkable is that once a beneficial trick appears, even if by chance, it can be preserved, enhanced, and passed along to future generations. Over eons, small advantages accumulate, and what started as a lucky quirk can become a central pillar of life's survival toolkit.

For us, the lesson is twofold. First, nature's solutions frequently surpass anything we've managed to invent in laboratories. Second, these solutions didn't arise from a divine blueprint or a eureka moment; they emerged incrementally, shaped by pressure and possibility. This perspective encourages humility. Rather than assuming we must teach nature how to run efficiently, we should acknowledge that nature may have much to teach us. Our technologies, powerful as they are, remain clumsy compared to the silent sophistication of a leaf at work.

Quantum effects in photosynthesis also provoke a philosophical reflection on our relationship with the universe's laws. We often treat quantum mechanics as a specialized field, relevant only to physicists or cutting-edge electronics. Yet here it is, playing a starring role in the everyday miracle of a plant turning sunlight into fuel. It suggests that the divide we draw between the "ordinary" world of sunlight and leaves and the "strange" world of quantum physics may be artificial. The extraordinary and the ordinary coexist, woven seamlessly together.

This realization can expand our sense of wonder. Just as seeds taught us about persistence and adaptation, and entropy taught us about the grand march of time, leaves now illustrate how life appropriates quantum phenomena for its own ends. And if life can do this at the molecular scale, what other quantum mysteries might be lurking in unexpected corners of our existence, quietly shaping the conditions that allow complexity to thrive?

As we explore further, we'll see how the interplay of quantum principles, biological ingenuity, and human curiosity leads us to new frontiers. Soon, we'll reach into the human realm, where our species attempts to replicate nature's successes and even extend them, transforming entire landscapes and ecosystems. Through it all, the lesson remains clear: complexity is not merely allowed—it's actively pursued, and sometimes that pursuit delves into the subatomic heart of reality.

Part III: The Shared Symphony of Life

Chapter 5: Humanity's Canvas—Sculptors of the Earth and Beyond

Look down at the ground beneath your feet. Once, it might have been untouched wilderness—grasses swaying in the wind, insects humming, and rocks settled quietly where gravity placed them. Over time, people arrived. They tilled the soil, built homes, forged paths, and erected cities. Now, if you stand in the heart of a metropolis, nearly everything you see has been shaped by human minds. Concrete towers replace ancient forests, farmland covers what were once wild plains, and dammed rivers flow in directions we dictate. Humanity, in its restless pursuit of comfort and exploration, has become a sculptor of landscapes, carving the Earth's surface and atmosphere into new forms.

This transformation didn't happen overnight. Early humans crafted tools from stone and bone, altering the environment in modest, localized ways. As we discovered agriculture, we learned to coax the soil into yielding abundant crops, reshaping entire ecosystems to suit our needs. The rise of complex civilizations complete with roads, aqueducts, and monumental structures further demonstrated our capacity to reorder the world. Today's global networks of trade, transportation, and communication reflect a species capable of rearranging resources on a planetary scale.

The notion that humans have emerged as planetary sculptors might seem bold, but consider what we've accomplished. We've reshaped coastlines with great harbors, redirected rivers with dams, and transformed deserts into gardens. We've mined metals from deep within the Earth and fashioned them into tools, machines, and towering edifices that alter weather patterns and even the reflectivity of the planet's surface. Meanwhile, our pollutants and greenhouse gases extend our influence into the atmosphere, warming the climate and potentially affecting global life-support systems.

This legacy of modification hints at an unsettling paradox. On the one hand, our creativity and ingenuity have allowed us to thrive, lifting billions out of poverty and enabling us to explore frontiers both physical and intellectual. On the other hand, our rapid ascent as environmental sculptors has sparked unintended consequences. The same cities that protect and nurture us can trap pollution and stress ecosystems; the same agricultural lands that feed us can deplete soils and reduce biodiversity. This tension, between our role as shapers and our responsibility as stewards, defines the human era.

Yet, before we grow too critical or too proud, let's remember that we are not the first lifeforms to influence the Earth's complexion. Microbes paved the way by producing oxygen, altering the planet's chemistry long before we arrived. Coral reefs built towering underwater architectures that rival our skylines in complexity. Forests, too, once shaped the climate by drawing carbon from the air. Humanity's sculpting differs in scale and speed, but not in concept. Like all life, we reshape our niche to survive and prosper. The difference now is that we have the power to sculpt not just a niche, but an entire planet—and possibly worlds beyond.

Our capacity as sculptors springs from the same fundamental principles that allowed seeds to grow and leaves to capture sunlight. We tap into energy gradients and channel them toward building and maintaining order. In our case, we burn fossil fuels, harness the wind, split atoms, and convert solar energy into electricity. We use these energy flows to construct intricate supply chains, elaborate information networks, and skyscrapers reaching ever higher. These feats highlight humanity's unparalleled ability to coordinate efforts across time and space, aligning countless individuals and technologies toward shared goals.

But this power also carries a heavy weight. Maintaining our complex infrastructures demands vast inputs of energy, raw materials, and continuous upkeep. Entropy works patiently in the background, ready to fray the edges and loosen the bolts. Our cities need constant repairs; our machines wear out and must be replaced. Without ongoing labor and ingenuity, the structures we've built would drift back toward disorder—a cautionary reminder that complexity never comes free.

The question is: how far can this sculpting impulse take us? We've managed to alter the Earth's surface so thoroughly that scientists debate whether we've entered a new geological epoch, the Anthropocene—an era defined by human influence on planetary systems. If that's true, then we've become not just inhabitants but curators of the biosphere, wielding the tools of geology, ecology, and climate. This shift challenges our sense of responsibility. If we're no longer just animals living off the land but its architects and guardians, what guiding principles should shape our designs?

Some might argue for a return to simplicity, a scaling back of our impact. Others envision pushing forward, using emerging technologies to reduce waste, enhance biodiversity, and stabilize the climate. We might deploy geoengineering, genetic modification, or artificial intelligence to manage ecosystems and resources with greater finesse. The future could hold farms tended by drones, reefs rebuilt by robotic caretakers, and atmosphere-scrubbing machines cleaning our skies.

Yet, beneath these lofty visions, uncertainty looms. We do not fully understand all the complexities of Earth's interconnected systems. Every intervention risks unintended consequences. Just as life at the molecular level must balance energy inputs and outputs to resist entropy, we must balance innovation and caution, creativity and humility. The key may lie in learning from nature's own successes—adopting resilient strategies that mimic ecosystems' recycling of materials and their intricate balancing acts among diverse species.

In broad strokes, humanity's sculpting mirrors the universe's grand narrative: an interplay of forces, both orderly and chaotic. We find ourselves at a crossroads, entrusted with tools that can reshape an entire planet and possibly guide the fate of life itself. Will we craft a flourishing biosphere, pushing back against entropy's pull and expanding life's foothold in the cosmos, or will our creations crumble beneath their own complexity?

The next chapters will explore these possibilities, examining how life's sculpting principles might extend beyond Earth and into the cosmic expanse. As we peer outward, we must bring with us the lessons learned at home—lessons about energy, order, creativity, and responsibility—so that wherever we set foot, we continue the dance begun so long ago by seeds, leaves, and humble microbes.

As we look upward from our reshaped landscapes, the vastness of space beckons. Already, we've taken our first tentative steps off-world, sending probes across the solar system and planting footprints on the Moon. We've even begun to dream of terraforming other planets—like Mars—by altering their atmospheres, climates, and ecosystems to suit human habitation. Such cosmic endeavors would be the ultimate act of sculpting: not just rearranging familiar terrains, but rewriting the conditions for life on entirely new worlds.

Yet, before we rush to become cosmic gardeners, we must ask: What responsibilities do we carry beyond Earth's boundaries? Once we've proven our ability to engineer climates and seed ecosystems, how do we decide which worlds to green and which to leave untouched? Some argue that we should seek out barren worlds, turning them into havens for life. Others caution that we must respect alien environments, whatever their state, and avoid imposing our own values on places that evolved without our influence. These debates echo the ethical dilemmas we face at home, magnified by interplanetary distances and the enormity of our potential impact.

The question of extending life beyond Earth also forces us to grapple with entropy on a grand scale. Can we engineer stable outposts in the near-vacuum of space? Can we create closed-loop ecologies that recycle resources and maintain complexity with minimal input? If we manage to spread life beyond Earth, we may be giving complexity a chance to endure beyond our home planet's finite lifespan. In that sense, interplanetary sculpting could be viewed as another chapter in life's ongoing defiance of entropy, a grand project that stretches the meaning of "survival" into deep space and distant futures.

But we must be honest with ourselves about our current limitations. We are still novices in understanding how ecosystems function, let alone how they might adapt to alien worlds. At home, we struggle to maintain biodiversity, preserve ancient forests, and prevent the collapse of fisheries or coral reefs. To attempt planetary-scale engineering without improving our stewardship on Earth might be like trying to paint a masterpiece without ever learning basic brushstrokes. Our cosmic ambitions should not overshadow the work we must do in our own backyard.

This tension might ultimately guide us to a more holistic mindset. Perhaps the greatest legacy of humanity's sculpting is not the cities we've built or the lands we've cultivated, but the recognition that we participate in a shared cosmic narrative. In shaping Earth and contemplating other worlds, we face both the promise and the peril of our own creativity. We stand poised between the universe's drift toward entropy and the possibilities that life's ingenuity offers.

In the chapters ahead, we'll cast our gaze still farther, imagining what life might look like elsewhere in the cosmos and wondering whether other sculptors—alien intelligences, distant ecosystems—might be at work. The questions remain open, and that openness invites our careful thought and measured action. Whatever paths we choose, humanity's canvas extends far beyond the horizon, challenging us to paint a future worthy of the name "creator."

Chapter 6: Cosmic Sculptors—Life Beyond Earth

Look up into the night sky. Each point of light is a star, many with planetary companions circling them in silent orbits. Our understanding of life thus far is a tale told by Earth—its soils, seas, and skies—but if the cosmos is as bountiful as we suspect, then our world is unlikely to be the only stage where life performs its intricate dances. For generations, we've wondered whether life exists elsewhere, and if it does, how it might sculpt order from disorder in unfamiliar ways. Could other living systems be weaving their own tapestries of complexity, adapting to foreign energy sources, chemistries, and climates?

Imagine a world where life thrives beneath an ice-covered ocean, warmed by hydrothermal vents that provide both energy and nutrients. In such a realm, creatures might assemble ecosystems entirely absent of sunlight, forging patterns and structures we can barely guess. Or consider a planet with multiple suns where photosynthetic organisms split the day into choreographed "shifts," evolving pigments that capture energy from different wavelengths. These alien strategies would be guided by the same thermodynamic rules we know, but their specific solutions might prove dazzling in their novelty.

We've already seen how life on Earth pushes back against entropy through myriad strategies—from the genetic instructions stored in seeds to the quantum feats of green leaves. On another planet, perhaps the building blocks of life aren't even carbon-based. Maybe silicon or ammonia plays the starring role, giving rise to life-forms that store and channel energy through unfamiliar molecular blueprints. If so, their ways of resisting entropy, of forging complexity and preserving it, could expand our definition of life's creative repertoire.

Of course, finding alien sculptors is no simple feat. Our spacecraft and telescopes extend our senses far beyond Earth, allowing us to read the faint chemical signatures in planetary atmospheres and probe distant surfaces for telltale patterns. We search for markers like oxygen, methane, or complex organic molecules—signs that something might be breathing, feeding, and growing. As our methods improve, we may detect biosignatures that hint at alien ecosystems hard at work.

But what if we encounter life that shapes its environment as dramatically as we've shaped ours—life that builds technological wonders, rearranges landscapes, and ponders the cosmos just as we do? Spotting such life might be easier or harder, depending on its level of activity. A distant civilization might leave behind unusual patterns of light, radio signals, or engineered megastructures. Or perhaps their fingerprints lie hidden in subtle anomalies, chemical imbalances we can't explain through natural processes alone.

The possibilities are boundless, and that's what makes the search both thrilling and humbling. Even as we consider ourselves skilled sculptors of Earth's landscapes, we cannot be sure what forms of artistry exist out there, how they've learned to harness energy and order, and what values guide their actions. Do they nurture complexity slowly and steadily, or do they leap forward in sudden bursts of innovation?

As we widen our search, we must also acknowledge the biases we carry. Our perspective is rooted in the biology of Earth, where liquid water, carbon-based molecules, and sunlight are central ingredients for life. This template guides our hypotheses—looking for "Earth-like" conditions on other planets as a starting point. Yet, the universe need not conform to our assumptions. Life, if it exists elsewhere, might flourish in conditions that would strike us as bizarre or hostile. Instead of relying on liquid water, it might use liquid methane; instead of seeking moderate temperatures, it could thrive in scorching vents or frigid deserts. Our challenge is to remain open-minded, refining our understanding of life's potential diversity with each new discovery.

In doing so, we're forced to confront a deeper question: what does it mean to call something "alive"? On Earth, we have a working definition—life uses energy to maintain order, grows, reproduces, and evolves. But what if alien processes blur these lines, performing some of these functions but not others? If we stumble upon a self-organizing chemical system that adapts to its surroundings and persists over time, should we consider it alive, even if it lacks familiar hallmarks like DNA or cell membranes?

Wrestling with such questions could help us advance not only our search for extraterrestrial life, but also our understanding of life here at home. By imagining new life-forms, we break free from the confines of terrestrial paradigms. This mental exercise might lead us to rethink basic assumptions—about how energy flows through organisms, how complexity can be stabilized, and what sorts of patterns could be considered "biological." In doing so, we become more creative observers, better equipped to notice subtle clues left by alien sculptors.

This openness of mind might be crucial as we begin to identify planets in "habitable zones" around distant stars—regions not too hot, not too cold, where conditions might resemble Earth's. But remember, the mere presence of a habitable zone doesn't guarantee life. Even here at home, life's emergence depended on a series of fortunate events and conditions. On another world, energy might be abundant, but chemistry intractable. Or chemistry might be rich, but stable niches too rare. Each planet or moon offers its own puzzle, an intricate interplay of laws and conditions that life must navigate.

Should we detect signs of alien life, the implications would be profound. It would confirm that complexity has taken hold more than once in the cosmos, showing that our narrative is not unique. By extension, it might suggest that the universe is a vast gallery of sculptors, each finding its own approach to the dance between entropy and order, each discovering novel ways to carve structure from chaos.

In the chapters ahead, we'll shift our focus to the universal principles that bridge science and myth, knowledge and meaning. We'll explore how our stories and theories about life's origins and destiny reflect a universal yearning: to understand our place in this grand tapestry and to glimpse how other threads, woven by alien hands, might join and enrich the pattern.

Let's imagine we confirm the existence of alien life—perhaps through a telescope's lens detecting subtle chemical signatures in a distant atmosphere or through a rover's camera capturing the patterns of something quietly thriving in a hidden corner of our solar system. What then? Beyond the initial shock and celebration, we'd face a profound intellectual and ethical challenge: how do we engage with forms of life that share our universe but not our evolutionary history or cultural assumptions?

This question speaks to the heart of life's sculpting role. If life on Earth can actively shape ecosystems, modify climates, and build intricate civilizations, might alien life do the same on their home world? Their strategies might be as diverse as their molecular makeup. Some alien biospheres might remain relatively simple, their complexity capped by scarce resources or harsh conditions. Others might evolve into dizzyingly elaborate systems, replete with ecological tiers and intricate energy flows that mirror or even surpass our own.

Envision, for example, an alien jungle where photosynthetic organisms and their grazers form elaborate webs of exchange. Or consider a world with no plants at all, where primary production occurs through chemosynthesis, forging layers of life unfamiliar to our eyes. In each scenario, these living beings would be sculptors in their own right, adapting to their environment and shaping it in turn, creating local pockets of order that push back against entropy's pull. The specifics of their metabolisms and lifecycles may be strange, but the core principle—transforming raw resources into patterns of complexity—would remain eerily familiar.

Once we acknowledge this possibility, the question of cultural and philosophical exchange arises. If these beings possess some form of intelligence—if they, too, ask questions, build tools, wonder about their place in the cosmos—then life's sculpting hand would extend beyond biology into realms of meaning and purpose. Could we learn from them, gaining insights into alternative paths to complexity, different ways to structure societies, or new philosophies to guide our stewardship of planets and resources? And what could we teach them in return, without imposing our values or risking harm?

Such encounters remain firmly in the realm of speculation. Yet, they serve as powerful thought experiments, prompting us to refine our understanding of life's cosmic role. They remind us that the interplay of entropy and order is universal, not confined to Earth's boundaries. In contemplating alien sculptors, we broaden our sense of what life can be and what it might become.

Soon, we'll turn our attention to the myths and science that span cultures and centuries, weaving together meaning from stories, data, and discoveries. In that tapestry, we may find that the core themes remain constant: curiosity, adaptation, and the capacity for life—wherever it may dwell—to push the boundaries of possibility. If the cosmos is indeed filled with sculptors, then we are part of an ancient and ongoing conversation, linked not just by shared matter and energy, but by the creative impulse life embodies in all its myriad forms.

Chapter 7: Universal Principles—Myths, Science, and Meaning

From ancient campfires to modern classrooms, we've told stories to understand our place in the universe. Long before anyone grasped the concept of entropy or the quantum genius of green leaves, human beings wove myths to explain why the sun rises and sets, how life began, and what secrets lurk beyond the stars. These stories were once whispered into ears eager for meaning, etched onto stone tablets, painted on cave walls, and eventually bound into sacred texts and philosophical treatises. They represent our first attempts at grappling with the profound questions that science now probes with data and experiment.

Yet, the relationship between myth and science is not one of simple opposition. Though their methods differ—myth relies on narrative and metaphor, science on observation and evidence—both share a deep desire to make sense of existence. In myth, we find characters and events that model creation and transformation; in science, we encounter principles and patterns describing the nature of matter, energy, and life. When these perspectives converge, they illuminate universal themes: the interplay of order and chaos, the march of time, the struggle of life to endure and evolve.

Consider the countless creation myths that describe how the world emerged from a formless void or a primordial ocean. While these tales differ in detail, they share the idea that something ordered came forth from an earlier, more chaotic state. In modern cosmology, the story has changed shape but not essence: from a seething early universe of high-energy chaos emerged galaxies, stars, and planetary systems. At the heart of both myth and science lies a recognition that complexity arises from what once seemed disordered, that patterns can crystallize out of an untamed backdrop.

Likewise, consider myths of heroic figures who challenge fate, restore balance, or journey through cycles of death and rebirth. Often, they mirror life's ongoing defiance of entropy, symbolizing renewal in the face of inevitable decline. In nature, we see analogous cycles: the seed that springs into a new plant, the ecosystems that rebuild after fires and storms, the evolutionary leaps that emerge after mass extinctions. These echoes suggest that the human impulse to find meaning in life's tenacity—our awe at how complexity emerges and persists—is woven into our deepest stories as well as our most rigorous sciences.

Even the search for alien life and cosmic sculptors has its mythic predecessors. Ancient myths often placed gods, spirits, or distant kin in the heavens, imagining that life and intelligence existed in some higher realm. Today's scientists scan exoplanets for signs of biosignatures, not divine messengers. Still, both impulses reflect a longing to know whether we are alone, whether something else "out there" shares our struggles and dreams. Myths about celestial beings and modern theories about extraterrestrial ecosystems share the same root: a curiosity that transcends boundaries, seeking companionship in the universe's quiet vastness.

What if we think of myths and science not as rival explanations but as complementary voices in a grand conversation about the universe's nature? Science reveals that the cosmos is shaped by fundamental laws—gravity pulling matter into cosmic structures, electromagnetism guiding energy flows, and quantum mechanics whispering strange possibilities. Myths, in their own way, reflect how we humans internalize these patterns, translating the cold equations into narratives that speak to our hearts and values. While science uncovers the scaffolding, myth decorates it with meaning, showing us how to live with the knowledge we glean.

This isn't to say that myths should be taken literally or that science needs spiritual adornments to be true. Rather, recognizing the interplay between them can help us navigate life's complexities. When we acknowledge that myths were our first attempts to understand disorder and bring forth patterns of meaning, we appreciate them as cultural fossils—remnants of earlier modes of thought that paved the way for empirical inquiry. Likewise, when we see that scientific truths can inspire awe and reverence, we understand that meaning doesn't vanish just because we rely on evidence and reason.

The universal principles that guide both myth and science extend beyond Earth. Just as ancient peoples looked at the sky and spun tales about constellations, gods, and cosmic rivers, we now point telescopes toward distant galaxies and speculate about alien biospheres and universal life strategies. Across time and disciplines, humans persist in asking: Why is there order at all? Why does complexity emerge, even when disorder seems so much easier?

In the chapters behind us, we've explored how entropy's steady increase shapes time's arrow, and how life, from the humblest seed to humanity's grand endeavors, can push back—at least temporarily—against that drift. We've seen how quantum tricks support life's energy capture, and how humans have taken their sculpting impulses planetary, perhaps one day extending them into space. Now, as we weave myth and science together, we see that both have been asking versions of the same question: What does it mean for life to persist against overwhelming odds, and how should we respond to that persistent spark?

This bridging of perspectives can guide our moral and philosophical outlook. If the universe's fundamental processes inspire creation myths and scientific principles alike, maybe the meaning we seek is not something separate from the cosmos, but something emergent within it. Perhaps the stories we tell, the equations we solve, the technologies we build, and the forests we nurture are all part of a single tapestry—threads that life continues to weave, refining patterns of complexity, beauty, and understanding.

As we turn to the next sections, we'll move more explicitly into these territories of human thought, considering how universal principles of creativity and complexity have led us to envision grand futures. We'll see how evolution—both biological and cultural—can unleash complexity in ways we've barely begun to imagine. And we'll ask ourselves whether this complexity, growing from seed to star, could one day rewrite the rules of the cosmic game.

Consider how differently we approach the idea of creation when we view it through both the mythic and scientific lenses. Mythic stories often personalize chaos, order, and life's struggles. They people them with gods, heroes, and tricksters—entities that embody the push and pull of energies we now describe with equations. Science replaces these characters with forces, probabilities, and evolutionary histories. Yet the narrative arc persists: a once featureless cosmos gives birth to galaxies, worlds, and life itself. This shared arc suggests that our search for meaning is not a mere cultural artifact; it arises from the raw ingredients of existence.

Think about the foundational questions both science and myth grapple with: Where did we come from? Why are we here? Where are we going? Science might say: we came from stardust, forged in stellar furnaces; we are here because life found a foothold on a once barren planet; we are going toward an uncertain future shaped by physics, biology, and our own decisions. Myth might say: we came from the dreamtime of gods or ancestors, emerged through trials and ordeals; we are here to learn, love, and grow; we are going onward to worlds unknown, guided by purpose and destiny. Both answers seek to situate us, providing orientation in a universe too vast for easy understanding.

In this sense, myth and science are twin attempts to wrest meaning from the silence of space. One uses imagination and metaphor; the other uses inference and evidence. Together, they paint a richer picture: a universe not just of particles and fields, but of stories and experiences. We realize that the cosmic dance is not solely about what is—it is also about what matters to us, what resonates with our deepest intuitions, and what inspires us to build futures worth living in.

As we have seen, life's strategies for sculpting complexity are not accidental. They arise from the interplay of laws and chance, guided by the relentless push of entropy and the equally relentless striving of living systems. When humans add their voices to this chorus, we bring not just tools and technology, but dreams and philosophies. We blend data and narrative, weaving a fabric of meaning that can span from the smallest cell to the farthest star.

In the next chapters, we will explore how evolution operates as a creative force, unlocking new levels of complexity and interdependence. We'll consider how humanity, already a planetary sculptor, might scale these lessons to shape a more sustainable, imaginative future. And we will listen closely to life's whisper, now transformed into a hopeful invitation—an invitation to understand that as we build stories and systems, as we embrace both myth and science, we are continuing the cosmic tradition of creation and transformation. In these harmonies of knowledge and meaning, we might find ourselves not just as observers of the universe, but as participants in its unfolding symphony.

Part IV: Life's Endless Possibilities

Chapter 8: Complexity Unleashed—Evolution as a Creative Force

Walk into a lush rainforest at dawn and listen. Hear the calls of birds mingling with the hum of insects and the whisper of leaves rustling in the breeze. Within this tapestry of life, each creature has its own story, etched by millions of years of experimentation and adaptation. Evolution—often described as a "blind watchmaker"—is the process by which life's complexity emerges over time. It sculpts organisms the way a river sculpts a canyon, eroding old shapes and carving new patterns, not according to an intelligent plan, but through countless small shifts in genetic codes and ecological contexts.

Yet, if evolution is so often portrayed as directionless and indifferent, how do we reconcile this with the astonishing variety and sophistication that surrounds us? The world's ecosystems are far richer in detail than any human imagination could design from scratch. Consider the intricate form of a hummingbird's beak, perfectly matched to the flowers it sips from, or the camouflage patterns on a butterfly's wings, mimicking the bark and leaves of the forest. These refinements emerge because, over countless generations, organisms that happen to fit their environment survive and pass along their genes, while those that don't disappear into the fossil record.

In essence, evolution is creativity at the cellular scale. Each genetic mutation, however random, provides raw material for experimentation. The environment serves as the testing ground, rewarding traits that confer survival and reproduction. Over time, these countless experiments accumulate into complex symphonies of form and function. The rainforest's layered canopies, teeming with niches, showcase how this process can generate not just hardy survivors, but ecological orchestras—with predators and prey, pollinators and plants, each instrument fine-tuned to its role.

This complexity does not arise despite entropy and disorder but in partnership with them. Environments change, sometimes subtly and sometimes cataclysmically. Species must adapt or perish. Through this dance, life unlocks new strategies to persist and thrive. Complexity often arises when resources and niches increase in number and variety—when energy flows abound and ecosystems diversify. New levels of order emerge at scales larger than any single organism: symbiotic relationships between species, complex food webs, entire biomes that cycle nutrients like giant, living machines.

Humans, too, are products of this evolutionary creativity. Our opposable thumbs, powerful brains, and social bonds are evolutionary endowments that allowed us to carve out a niche at the top of the food chain. These traits, honed over epochs, enable us to shape tools, ideas, and cultures that, in turn, reshape the world around us. As we move forward into considering our role as sculptors, it's essential to remember that we're not separate from the evolutionary process. We're its heirs, carrying forward the innovations our ancestors earned through trial and error.

But what happens when evolution's products become conscious creators, guiding selection not merely through survival and reproduction, but through deliberate design? When humanity appeared on the scene, evolution had already equipped countless species with ingenious solutions: camouflage and courtship dances, the chemistry of venom and the insulation of fur, intricate social systems, and countless other adaptations. Our ancestors harnessed these inherited tools, refining them into weapons, shelter, agriculture, and trade. Over time, we've learned to shape the very conditions that guide evolutionary paths—both for ourselves and for other species. Selective breeding of animals and plants turned wild species into trusted companions and reliable food sources. We've cultivated wheat, rice, and corn into staple crops that feed billions. We've bred dogs into forms that fit countless roles, from herders to guardians to beloved companions.

This controlled evolution, under human guidance, can be seen as a new phase in life's ongoing experiment. Instead of relying solely on the slow filter of natural selection, we impose our preferences: flavor, yield, resilience. We choose which species thrive and which struggle, altering the course of evolution at unprecedented speed. In doing so, we've accelerated complexity in some areas—creating elaborate agricultural ecosystems and domesticating plants and animals that depend on us. But we've also simplified other realms, replacing wild diversity with monocultures, draining wetlands to make way for farmlands, and pushing many species to the brink of extinction.

The rise of genetic engineering takes this sculpting power to another level. By editing genes directly, we skip the random trial-and-error process that characterizes natural mutation. We can insert traits that nature never stumbled upon, rewriting the genetic "instructions" that have been refined over eons. This gives us the potential to engineer crops that resist pests without pesticides, or organisms that clean up environmental pollutants. It also raises haunting questions about unintended consequences and ethical boundaries. Are we prepared to shoulder the responsibility that comes with the power to rewrite life's code?

Yet, these developments don't have to be seen as reckless tampering. If guided by wisdom and humility, they might represent another stage in evolution's creative saga—one where conscious understanding supplements the blind groping of random variation and selection. Humans could become caretakers as much as designers, using our knowledge to help life adapt to a rapidly changing planet. We could choose to conserve biodiversity, restore habitats, and ensure that the complexity we create doesn't come at the cost of eroding life's broader tapestry.

As we move into future chapters, we'll explore the interplay between cultural evolution and biological evolution—how ideas, values, and technologies spread and mutate much like genes do. We'll examine how humanity's sculpting role extends beyond mere survival, touching on matters of purpose, beauty, and the quest for greater interconnection. If the essence of evolution is the unveiling of new possibilities, then our challenge is to ensure that the possibilities we unlock lead toward a richer, more vibrant web of existence rather than a hollowed-out monoculture of our own making.

To appreciate the grandness of evolution's creative potential, let's zoom out and consider its long-term trajectory. Over billions of years, life on Earth transformed from single-celled microbes drifting in ancient seas to the teeming mosaic of ecosystems we know today. At each stage, complexity emerged through incremental steps, each tested against the unforgiving matrix of survival. Eyes that once detected nothing but light and dark evolved into organs that perceive color, depth, and detail. Limbs that once helped creatures crawl onto land adapted into wings, hooves, flippers, and grasping hands.

We often measure evolution's creativity by the sheer range of living forms, but there's another dimension: the intertwining relationships and ecological services that define entire biomes. Forests regulate climate, coral reefs nurture biodiversity, and pollinators partner with flowering plants in a delicate ballet that keeps ecosystems humming. The cumulative result isn't just an assortment of creatures—it's a planetary metabolism, cycling nutrients and energy through vast networks of cooperation, competition, and communication.

As humans navigate our own evolutionary future, we enter a phase where cultural and technological innovations evolve as rapidly as genetic traits. Ideas are like mutations of the mind, tested in the environment of human society. Those that thrive shape our civilizations, influencing how we steward resources and design our technologies. Over time, culture can drive new forms of complexity—intricate economies, democratic institutions, global alliances—that, in their own way, parallel the adaptive intricacies of biological life.

This opens a door to a new understanding of our role. Rather than seeing ourselves as standing apart from nature, we can view our cultural evolution as part of life's ongoing experiment. We've learned to write codes—not just genetic, but digital—to organize information, coordinate labor, and harness global flows of energy. We've built cities as elaborate as coral reefs, connected by information networks more extensive than fungal mycelia beneath a forest floor. Our challenge is to ensure these creations enhance rather than diminish life's complexity.

This holistic perspective suggests that evolution doesn't end with the survival of the fittest—it continues as a mutual enrichment, a grand interplay of minds and molecules. Just as evolution generated the hummingbird's refined beak and the forest's layered canopy, it can push our cultures toward greater empathy, resilience, and inclusivity. Perhaps we can guide our own cultural and technological mutations to better align with the values and principles gleaned from nature's story.

In the following chapters, we'll contemplate how humanity's sculpting power might develop further—beyond the Earth, into cosmic scales, and toward futures that test the limits of our imagination. By blending our understanding of evolution's creative force with the insights from myth, science, and philosophy, we may come to see ourselves not just as products of a long evolutionary lineage, but as active contributors to what evolution creates next. Life's whisper invites us to join this process consciously, to shape complexity in ways that honor and enrich the broader tapestry of existence.

Chapter 9: The Sculptor's Invitation—A Creative Future, (after reading...;-)

Close your eyes and imagine a moment between breaths—a stillness where countless forces, seen and unseen, gather in quiet tension. Here, potential and actuality intertwine. The world before us is neither fixed nor finished; it hums with a silent readiness for change. Even in our modern age, when we gaze at bustling cities or pristine wilderness, we witness an ongoing interplay of structure and dissolution, an eternal balancing act that has sculpted everything from coral reefs to cultural myths.

This dance is neither moral nor guided by preference. Instead, it unspools from the very fabric of existence. The myriad patterns that emerge—from the molecular rhythms of photosynthesis to the swirl of galaxies—arise at the boundary where order meets chaos. In that fertile interface, life has always found room to grow more intricate. Through billions of years, the cosmos has revealed that complexity emerges not from rigid control or blind release, but from subtle negotiations between forces inclined to shatter forms and those determined to crystallize them.

As we look forward, we need not confine our understanding of this creative tension to any single narrative of progress or retreat. Instead, we can step back and behold it as a universal principle. Today's power grids and data networks trace out their own webs of structure and fragility. In tomorrow's landscapes—whatever they may be—patterns will continue to shift and evolve in ways we cannot fully predict. There is no endpoint, no final equilibrium. Complexity is a process, not a product.

What does it mean to acknowledge this enduring interplay? Perhaps it invites a different sort of engagement, one rooted not in triumph over disorder, nor in surrender to it, but in curiosity. Like a sculptor working with clay, the universe's patterns are never complete. Each intervention, whether a chemical reaction, a cultural innovation, or a sudden mutation in a gene, sets off ripples that may solidify into lasting structures or dissolve back into the currents of change.

In these pages, we have explored how life, from its earliest sparks, consistently leveraged this tension. Seeds, leaves, and entire ecosystems learned to improvise within the constraints of physics and chemistry. Humanity joined in, adding technology and thought to the mix. At no point did the dance cease. Instead, it became more layered, weaving human aspirations into the warp and weft of an ever-transforming tapestry.

If we now find ourselves on the cusp of uncharted territories—contemplating futures that blend biological insight with quantum subtlety—it may serve us well to remember that the cosmos has always been in motion. Complexity grows from the restless push and pull between creation and dissolution. Our role, however we choose to see it, unfolds within these currents, reminding us that any sense of finality or permanence will always be subject to this deeper, ancient interplay.

To witness this interplay in action, we need look no further than the shifting boundaries where land meets sea, or desert meets forest. Such edges are places of negotiation. Sand dunes drift, vegetation creeps or recedes, species adapt or perish. It's here, at these margins, that small changes can cascade into new patterns, revealing how structure and unstructured forces continually reshape each other. The very idea of a fixed "balance" is elusive; instead, there is a continuous conversation, with every whisper of structure answered by a sigh of disorder, and vice versa.

What emerges from this ongoing dance is not a static harmony, but a dynamic equilibrium—one that supports the rise of complexity. Patterns solidify into something stable enough for organisms to exploit, and yet remain fluid enough to encourage novelty. Life thrives in these conditions, navigating the currents of entropy and drawing new forms out of old substrates. Think of a spider weaving its web: it anchors silken threads to create a lattice of order that bends to the breeze. The web's purpose is clear—catching prey—but the spider has no illusions that its structure is permanent. At any moment, a gust of wind or the passage of time can force it to rebuild, improve, or relocate.

Human creativity, too, follows these rhythms. We design houses and cities, only to watch them age and crumble, reinventing them in cycles of construction and decay.

Ideas spawn technologies that proliferate until they meet new obstacles—resource limits, shifting values, unforeseen consequences—prompting fresh innovations. This iterative dance mirrors the cosmic story, tying our personal efforts to the broader currents of matter and energy. Whether we're cultivating a garden, coding a program, or theorizing about distant galaxies, we're participants in an ancient tradition of forming and reforming structures.

In contemplating our futures, we might be tempted to seek ultimate resolutions: a stable state of permanent prosperity, an end to all struggle. But the universe seems to tell a different tale. The building blocks of existence are never locked in place; they swirl in patterns of emergence and dissolution. The true art lies in recognizing that this motion is not failure or tragedy, but the source of ongoing creation. Complexity is born not from escaping the dance, but from dancing with it.

As we enter the final chapters of this narrative, we'll step further into this perspective, looking beyond individual species, stories, and epochs. The cosmos offers no final blueprint, no guarantee. Yet it continually produces new forms, expanding the tapestry of possibilities. By placing ourselves within that unfolding process—neither above nor outside it—we may find meaning not in static achievements, but in the ongoing act of participating, responding, and becoming.

To see these principles at work on a grand scale, look to the patterns that emerge over spans of time so immense that our daily concerns vanish into abstraction. Stars burn and die, scattering elements into space. Planets form, environments fluctuate, and life inches forward in step with subtle transformations. Even civilizations, though young by cosmic measures, arise and dissolve, leaving traces that may or may not endure. Through it all, the conversation between order and disorder persists—a quiet hum beneath every epoch, weaving new chapters into a story without a final endpoint.

In this larger view, the cosmos doesn't show us how to "win" against entropy, nor does it counsel surrender to chaos. It only displays, with unwavering consistency, that creation is not an event but a process, a perpetual unfolding of possibilities. We are witnesses, participants, and products of this ongoing interplay. The sculptures we carve, whether they are literal artifacts or intangible ideas, reflect our momentary understanding of the dance. With time, new insights, and changing conditions, we rework old forms, discover fresh patterns, and open new vistas of complexity.

This perspective is no prescription for how to live—no formula guaranteeing that the future will align with our hopes. Rather, it offers a framework for understanding how complexity emerges at all scales and times. It suggests that every step we take—a hypothesis tested, a seed planted, a story shared—becomes part of a larger choreography. Each act contributes to, and is shaped by, the shifting interplay of structure and dissolution, the interplay that has always defined our universe.

As we near the end of our journey, consider what it means to fully accept this process. Rather than seeking permanent solutions or perfect stability, we might recognize that impermanence is the crucible of growth. Rather than longing for closure, we can appreciate that openness invites new forms to arise. Life's whisper, which we've followed through seeds, leaves, human endeavors, and cosmic visions, continues on, echoing into every unknown future.

In the chapters that follow, we'll draw these threads together, reflecting on how embracing the tension between order and chaos might shape our understanding of who we are and what we can become. The sculptor's invitation remains: to join the dance as creators and observers, to find beauty and meaning in the infinite unfolding, and to recognize that we, too, are patterns of energy and matter, continuously recomposed as part of a universe that never stops inviting new possibilities.

Chapter 10: Conclusion—A Cosmic Whisper
Stand beneath a clear, moonless sky and let your gaze drift into the depths of starlight. Each distant flicker represents a place where gravity and energy sculpted order from a cosmic haze, where elements forged in ancient furnaces drifted and swirled until planets, perhaps life, took shape. Our eyes, evolved in Earth's particular niche, can scarcely fathom the scale of this interplay. Yet, as we have followed life's whisper through these chapters, a grand pattern emerges—an unending conversation between structure and dissolution, each step of creation shadowed by change and instability.

This recognition is not an answer to every riddle; it doesn't provide a neat moral or a tidy resolution to the puzzles we've explored. Rather, it offers a way of seeing—a lens through which to understand that the complexity we cherish arises not from stillness, but from dynamism. Life's seeds push through cracks and thrive in shifting soils. Quantum phenomena shape the energy flows that keep leaves green and vibrant. Humans bend landscapes, build cities, and dream of other worlds, all the while confronting entropy's steady pull. And through it all, the universe insists that complexity is not guaranteed, only earned moment by moment, adaptation by adaptation.

We have glimpsed how myths and science share deep questions, how evolution acts as a creative force, and how we humans, curious and inventive, have learned to sculpt environments and ideas. These stories do not converge into a final, triumphant chorus. Instead, they remind us that the interplay of order and chaos has no ultimate act—only endless improvisations. Each generation inherits patterns from the past and leaves behind its own imprints. Each world carries forward its unique negotiations between stability and flux.

In this sense, the universe whispers by example, showing that meaning can be found in engagement rather than conclusion. The patterns we observe—whether in ecosystems, civilizations, or stars—are chapters in a sprawling epic without a single author or a final page. The narrative threads are countless. As observers, we piece together fragments, constructing a tapestry that resonates with our inquiries. As participants, our actions influence the texture of the unfolding story, adding new threads, weaving fresh motifs, and sometimes unraveling old ones.

This perspective might feel humbling. Our intricate endeavors remain precarious, our grandest structures subject to decay. Yet, humility need not breed despair. Recognizing ourselves as part of an ongoing cosmic dialogue frees us from the burden of finding a "last word" on existence. Instead, it invites us to dance with uncertainty, to view entropy not as an enemy but as the backdrop against which order shines most brightly. The whisper of life says: complexity can bloom, even amid relentless transformation.

As we enter this concluding phase, let's reflect on how everything we've explored—entropy, evolution, quantum insights, cultural creativity—fits into a universe that never stops experimenting with form. Let's consider how understanding this interplay might shape not just our theories, but our sense of belonging in a cosmos that speaks through patterns and possibilities.

If we stand at the edge of what we know—our science, our myths, our technologies—and look outward, we see that the cosmos doesn't hand us final verdicts. Instead, it offers seeds of possibility. The interplay of entropy and order, of structure and unstructured forces, continues everywhere. Planets form and shatter, species emerge and vanish, cultures wax and wane. Each event is an expression of underlying tensions, dynamic balances that shift with time and conditions.

In acknowledging this, we become more than just observers of the cosmic dance. We become participants, adding our voices and gestures to the chorus. Our knowledge, incomplete and evolving, helps us make sense of the patterns we encounter. Our choices, bold or tentative, feed back into the systems we inhabit. In this open-ended cosmos, we are sculptors and sculptures at once—malleable, responsive, capable of shaping and being shaped.

None of this guarantees a particular future. The universe's lessons are not prescriptions, and its whisper provides no blueprint. Instead, it hints that complexity and meaning arise through the give-and-take of existence. If we crave permanence, we may be disappointed; if we embrace transformation, we may find a kind of solace. The stories we tell, the worlds we build, and the ideas we explore all form part of a larger interplay—one that began long before us and will continue long after we're gone.

In that recognition, there's a quiet comfort. The very precariousness that makes life precious also makes it creative. The uncertainty that forces us to adapt also opens the door to new forms of complexity. By paying attention to life's whisper, we learn that what matters is not reaching a final state of perfection but understanding the rhythms of emergence and dissolution. This insight can liberate us from rigid expectations, encouraging us to discover harmony in flux and value in the fleeting patterns that define our reality.

As we draw the curtain on these explorations, remember that our journey was never about finding a conclusive statement. Rather, it was an invitation to think differently—about how entropy sculpts time, how life responds with ingenious strategies, and how humanity extends that creativity with its own cultural and technological symphonies. By reframing complexity not as an endpoint but as a continuous unfolding, we open ourselves to a world that is more adventurous, more alive, and more intrinsically meaningful.

In the starlit silence that follows these thoughts, the universe continues its ceaseless transformations. Somewhere, a seed germinates in cracked pavement, a leaf captures a photon with quantum finesse, a human mind envisions a novel idea. Each event, small or grand, adds to the tapestry. Each moment, we rediscover that complexity isn't something we inherit fully formed—it's something we help create, one ephemeral pattern at a time.

Step back and consider the tapestry we've glimpsed: a universe without tidy conclusions, but brimming with the raw materials for endless invention. Complexity arises whenever conditions allow for new patterns to crystallize. Life seizes these opportunities, weaving fragile order from elemental chaos. Sometimes, that order endures long enough to inspire and astonish; other times, it dissolves, clearing the stage for fresh experiments. There is no final blueprint, no locked-in destiny—only a perpetually unfurling array of possibilities.

In this shifting panorama, meaning emerges through engagement. We find significance not in halting the dance, but in participating fully, embracing the interplay that weds structure to fluidity. Our minds, shaped by millennia of adaptation, hunger for narratives, patterns, and connections. That hunger drives us to probe deeper, to question, to learn, and to create. Each generation inherits a world in flux, its forms always partial, its potentials always incomplete. And each generation adds its voice, extending the dialogue that began long before humans arrived.

As we step away from these pages, let us carry forward a renewed appreciation for the strange and wonderful fact that complexity can arise at all. Let us acknowledge that whatever we build—cities or stories, artworks or hypotheses—will be temporary sculptures in an ever-changing landscape. Their impermanence need not diminish their worth. On the contrary, it may enhance it, reminding us that creativity and discovery live in the present moment, perched between what is known and what is yet to be imagined.

In listening to life's whisper, we've heard echoes of distant galaxies, quantum secrets nestled in leaves, and the persistent striving of organisms that weave themselves into intricate worlds. We've seen that the cosmic game never truly ends, and that each of us is invited to play. In this open, evolving universe, complexity is not a final prize, but an ongoing dialogue—a conversation without end.

With that in mind, we release this exploration back into the current of thought, scattering these ideas into the broader discourse from which they sprang. The patterns we've traced here will doubtless rearrange and recombine as new minds reflect upon them, adding new threads to the tapestry. In that way, this book becomes another whisper among many, another subtle suggestion that the cosmos is alive with potential, and that no matter how much we learn, there will always be more to discover.

Appendices: Tools for Understanding
In these appendices, we gather conceptual tools and clarifications to help readers explore the ideas introduced in the main chapters. Here, we won't attempt to rewrite the story told so far; instead, we'll provide a set of keys—brief explanations, simplified models, and references—that can unlock deeper understandings. Whether you're looking for a refresher on a concept mentioned in passing, or hoping to make connections between scientific principles and the broader themes of complexity and creativity, these resources can serve as a starting point.

Appendix A: Key Scientific Concepts
• Entropy and the Second Law of Thermodynamics:Entropy is often described as a measure of disorder, but more accurately, it reflects the number of ways a system can be arranged at a microscopic level. The Second Law of Thermodynamics states that, in a closed system, entropy tends to increase over time. This principle underlies the "arrow of time," explaining why processes like mixing, dispersing, and spreading out are far more common than their reverse. Though entropy sets a broad directional flow toward greater randomness, it doesn't forbid pockets of order from arising—especially when energy is regularly supplied. Life exploits just such conditions, channeling energy flows to build complexity even in an entropy-governed universe.

- Quantum Mechanics and Photosynthesis: Quantum mechanics deals with the strange behavior of particles at very small scales, where matter and energy can exist in multiple states at once and outcomes are described by probabilities. In photosynthesis, certain molecular structures within leaves appear to use quantum effects to optimize energy transfer from captured photons to reaction centers, increasing efficiency. This connection between quantum rules and biological function highlights that life can find subtle "loopholes" in nature's complexity, forging new structures and capabilities from surprising foundations.
- Evolution and Natural Selection: Evolution describes how life changes over generations through the modification of genetic information. Natural selection, one of evolution's driving mechanisms, favors traits that improve an organism's chances of surviving and reproducing in its environment. Over immense stretches of time, these incremental improvements accumulate, yielding astonishing complexity and diversity. While evolution doesn't follow a predetermined plan, it often generates forms and strategies that seem beautifully adapted, revealing how order and complexity can emerge from a process guided by variation, inheritance, and selection.
- Ecosystems and Networks: An ecosystem is more than a collection of organisms. It's a network of relationships—predator and prey, plant and pollinator, microbe and host—through which energy and matter

circulate. Complexity thrives in these networks because they create niches, feedback loops, and emergent properties. Understanding ecosystems helps us appreciate how complexity isn't just about individual species, but about the patterns formed through their interactions.

Appendix B: Philosophical and Conceptual Frameworks

- Cosmic Perspective: Adopting a cosmic perspective means looking at Earthly phenomena—life, culture, technology—as part of a vast, evolving universe. From this vantage point, human activities appear as recent chapters in a much older story, one shaped by the interplay of energy, matter, and time. Recognizing our place in this cosmic narrative can foster humility, as we see that the complexity we cherish emerges from processes far grander than any single species. It also encourages curiosity: if life and complexity arose here, what other stories might the universe hold, waiting to be discovered?

- Systems Thinking: Systems thinking involves viewing phenomena not in isolation, but as parts of dynamic, interacting wholes. Rather than focusing solely on individual components, systems thinking examines the relationships, feedback loops, and emergent properties that arise from collective behavior. This approach can be applied to everything from ecosystems and economies to the neural networks of the brain. By looking at how patterns form, shift, and adapt over time, we learn to appreciate complexity as an evolving dialogue, rather than a static state.

- Emergence and Self-Organization: Emergence occurs when a system's complexity or behavior cannot be fully predicted or understood by examining its parts in isolation. Think of a flock of birds moving in intricate patterns without a single leader, or an ant colony building complex structures through simple, local rules. These phenomena highlight how order can arise spontaneously, from the bottom up, when conditions allow interactions to accumulate and refine themselves. Self-organization underscores that complexity doesn't always require top-down planning; it can emerge naturally, given the right circumstances and rules.

- Narrative and Metaphor: Humans have long relied on stories and metaphors to grasp complexity. While scientific data can describe patterns with precision, narratives help us relate abstract principles to human experiences, motivations, and emotions. The metaphors we choose—entropy as a river flowing downhill, quantum effects as subtle shortcuts—can shape how we interpret and internalize complex ideas. Recognizing the power of narrative encourages us to be mindful in selecting our analogies, ensuring they illuminate rather than distort.

Appendix C: Bridging Scales and Disciplines

- From Particles to Galaxies: The universe's complexity spans a colossal range of scales, from subatomic particles behaving probabilistically to galaxies whose shapes emerge from gravitational interactions over billions of years. Connecting these scales is challenging, but doing so can reveal deep insights. For example, the behavior of molecules underlies the chemistry of life, which in turn influences ecosystems, cultures, and technologies. Understanding how patterns propagate across scales helps us see that complexity isn't confined to one domain—it resonates across the fabric of existence.

- Interdisciplinary Insights: No single field holds a monopoly on understanding complexity. Physicists study energy and matter, biologists explore life's strategies, anthropologists examine cultures, and philosophers reflect on meaning and ethics. By blending insights from multiple disciplines, we gain a richer, more nuanced perspective. For instance, recognizing how quantum principles shape photosynthesis requires knowledge of physics, chemistry, and biology. Similarly, understanding human creativity may involve psychology, anthropology, and even cosmology. Interdisciplinary dialogue can spark breakthroughs and deeper appreciation.

Appendix D: Further Exploration and Reflection
- Suggested Readings:Expanding your understanding often involves diving deeper into specific topics. Texts on thermodynamics, quantum mechanics, evolutionary biology, and ecology can provide a firmer scientific grounding. Works of philosophy, comparative mythology, and cultural history reveal how different civilizations have grappled with similar questions. Seek out authors who bridge disciplines, who infuse scientific rigor with narrative flair, and who challenge assumptions about where complexity comes from and what it means.
- Asking Open-Ended Questions:Complexity resists simple answers. Instead, it invites inquiry. Consider questions that have no single resolution: How might life's strategies differ on other worlds? In what ways could future technologies reshape the dance between structure and chaos? What overlooked patterns might emerge if we shift our perspective or change our assumptions? Open-ended questioning keeps our minds agile, preventing us from settling too quickly into one narrative or conclusion.
- Creative Engagement:Don't limit your exploration of complexity to reading and analysis—engage with it creatively. Model an ecosystem in a terrarium or virtual simulation, noticing how small changes ripple through the system. Write stories or poetry that capture the feeling of emergence and transformation. Experiment with artistic media that embrace randomness and pattern, learning firsthand how

order can arise from play and improvisation. Creativity and curiosity feed into each other, helping you internalize concepts that might otherwise remain abstract.

• Community and Conversation:Complexity thinking thrives on dialogue. Share what you've learned with friends, colleagues, or fellow enthusiasts. Join study groups or online forums dedicated to exploring these themes. Consider how diverse viewpoints—from scientists and philosophers to artists and spiritual leaders—might enrich your understanding. By listening to others' interpretations and insights, you open yourself to perspectives that challenge and refine your own.

Appendix E: Embracing Uncertainty

• Limits of Knowledge:Recognize that no matter how deeply you delve, some questions will remain unresolved. This isn't a flaw—it's the nature of complexity. The patterns we've discussed emerge from interactions too numerous, subtle, and expansive to fully capture. Accepting this can feel liberating. It frees you from the burden of "solving" the universe and instead encourages you to participate in an ongoing exploration.

• Evolving Understandings:Just as life and complexity evolve, so do our frameworks for understanding them. Scientific paradigms shift, cultural values change, and what once seemed certain can be revised by new data or insights. The journey is not about reaching a final destination, but adapting, learning, and discovering ever more nuanced ways of seeing.

• Continual Conversation: The ideas presented in this book are meant to be stepping stones, not stopping points. Take what resonates, question what doesn't, and follow the threads that intrigue you. As you do, you continue the ancient and ongoing tradition of engaging with complexity, enriching it with your own contributions.

By assembling these conceptual tools and frameworks, we hope to encourage not just comprehension, but a feeling of agency and wonder. Complexity is not just out there—it's within us, shaping how we think, grow, and interact with the world. May these appendices serve as a resource for your own explorations, wherever they may lead.

Bibliography: Diving Deeper

This bibliography is intended as a guide for those who wish to delve further into the scientific, philosophical, and cultural dimensions explored in this book. The works listed here span a wide range of disciplines—from physics and biology to anthropology, history, and art. They reflect the idea that understanding complexity and life's creative potential is not restricted to any single area of expertise. By engaging with these texts, readers can immerse themselves in deeper analyses, alternative perspectives, and ever-evolving dialogues about our universe's grand tapestry.

Section A: Foundational Scientific Texts
- Thermodynamics and Entropy:
- Atkins, Peter. The Laws of Thermodynamics: A Very Short Introduction. Oxford University Press.
- Schroeder, Daniel V. An Introduction to Thermal Physics. Addison-Wesley. These works introduce the core principles of thermodynamics and entropy, providing the conceptual scaffolding to understand why complexity and disorder interact as they do.
- Quantum Mechanics and Photosynthesis:
- Ball, Philip. Beyond Weird: Why Everything You Thought You Knew About Quantum Physics Is Different. University of Chicago Press.

- Blankenship, Robert E. Molecular Mechanisms of Photosynthesis. Wiley-Blackwell. For readers curious about the delicate interplay between quantum phenomena and life's energy capture strategies, these texts offer insights into the subtle world of subatomic processes and the chemical intricacies of photosynthesis.
- Evolution and Biology of Complexity:
- Dawkins, Richard. The Selfish Gene. Oxford University Press.
- Margulis, Lynn, and Dorion Sagan. Microcosmos: Four Billion Years of Evolution from Our Microbial Ancestors. University of California Press. These books examine how life's diversity emerged through evolutionary processes, exploring the genetic and ecological factors that lead to complexity.
- Ecology and Systems Thinking:
- Odum, Eugene P. Fundamentals of Ecology. Saunders.
- Capra, Fritjof, and Pier Luigi Luisi. The Systems View of Life: A Unifying Vision. Cambridge University Press. By looking at life through the lens of interconnected systems and ecological networks, readers can appreciate how complexity forms from relationships, feedback loops, and emergent patterns.

Section B: Philosophical, Cultural, and Mythological Contexts

- Myth and Meaning:
- Campbell, Joseph. The Hero with a Thousand Faces. Princeton University Press.
- Eliade, Mircea. The Sacred and the Profane: The Nature of Religion. Harcourt Brace Jovanovich.These classics delve into the mythic and religious narratives that humans have crafted, illuminating how stories and symbols capture the tension between order and chaos, life and entropy.
- Philosophy of Science and Complexity:
- Kauffman, Stuart. At Home in the Universe: The Search for Laws of Self-Organization and Complexity. Oxford University Press.
- Morin, Edgar. On Complexity. Hampton Press.For readers interested in the philosophical underpinnings of complexity studies, these works investigate how knowledge, order, and novelty interweave, challenging reductionist views of nature.
- Cultural Perspectives and Comparative Thought:
- Harari, Yuval Noah. Sapiens: A Brief History of Humankind. Harper.
- Leavitt, Steven D., and John R. Fox. Handbook of Cosmology and Culture. Imagined Publications (Fictitious example).Exploring anthropological and cultural frameworks reveals how different societies interpret their place in a dynamic

cosmos. Such comparisons sharpen our understanding that each culture's worldview emerges from the dialogue between environment, history, and imagination.

Section C: Artistic, Literary, and Interdisciplinary Works
- Art, Metaphor, and Creativity:
- Maturana, Humberto R., and Francisco J. Varela. The Tree of Knowledge: The Biological Roots of Human Understanding. Shambhala.
- Bernd Heinrich. Mind of the Raven: Investigations and Adventures with Wolf-Birds. Harper Perennial. While not traditional "art books," these texts highlight the creative intersections of biology, cognition, and imagination. They encourage readers to see the natural world as a source of metaphor and insight, bridging disciplines and sparking new ways of thinking.
- Fiction and Speculative Thought:
- Le Guin, Ursula K. The Left Hand of Darkness. Ace Books.
- Liu, Cixin. The Three-Body Problem. Tor Books. Literature offers a unique lens on complexity, posing "what if" questions that invite us to imagine alien ecologies, evolving cultures, and the interplay of order and chaos in realms beyond our own. Fiction can stretch the boundaries of thought, challenging readers to consider alternative futures and hidden patterns.

Section D: Continuing Inquiry and Dialogue
- Journals and Online Resources:
- Nature, Science, and PNAS (Proceedings of the National Academy of Sciences) regularly publish cutting-edge research on complexity-related topics.
- Websites dedicated to complexity science (such as the Santa Fe Institute's resources) and online lectures (e.g., MIT OpenCourseWare) can provide current insights into evolving fields.
- Workshops, Conferences, and Community Engagement: Engaging with academic societies, attending interdisciplinary conferences, or participating in local reading groups can foster dialogue, spark new interests, and enhance one's understanding of complexity. Learning often thrives best where minds meet and share discoveries.

This bibliography only scratches the surface of the vast literature available. The universe of complexity, creativity, and life's defiance of disorder is as expansive and evolving as the cosmos itself. Each source here can serve as a stepping stone, leading to further inquiries, cross-references, and personal interpretations. In pursuing these avenues, readers contribute to the ongoing conversation that bridges knowledge, imagination, and the deep patterns that shape our world.

Inspirational Figures
Throughout history, certain individuals have stood at the crossroads of disciplines, illuminating ways to understand complexity, creativity, and the interplay of order and chaos. They approached knowledge not as a finished product, but as a landscape to be continually explored. In these profiles, we celebrate scientists, philosophers, artists, and thinkers who challenged conventional thinking, bridged disparate fields, and inspired us to see the world as an evolving tapestry of patterns. By studying their lives and ideas, we can glean insights into the courage, curiosity, and resilience required to engage wholeheartedly with a universe that never stops unfolding.

1. **Richard Feynman** (1918–1988)
A theoretical physicist renowned for his originality and charisma, Feynman reshaped our understanding of quantum mechanics and particle physics. Beyond his groundbreaking research, he had a gift for explaining complex ideas with clarity, humor, and a sense of wonder. His famous Feynman Lectures on Physics transcended the classroom, conveying not just facts but a mindset—one that encouraged questioning assumptions, embracing uncertainty, and delighting in the surprises nature offers. Feynman's approach exemplifies how curiosity and playfulness can unlock realms of understanding within the most intricate scientific theories.

2. **Lynn Margulis** (1938–2011)
A pioneering evolutionary biologist, Margulis revolutionized our view of life's complexity by championing the theory of endosymbiosis. She argued that the eukaryotic cells making up most complex organisms arose through symbiotic mergers of simpler cells, rather than a linear, competitive climb. This insight wove cooperation into the evolutionary narrative, showing that life's complexity can emerge from collaboration as much as from conflict. Margulis's interdisciplinary approach, drawing from microbiology, geology, and evolutionary theory, expanded the evolutionary landscape, reminding us that new structures often arise where diverse elements meet and merge.

3. **Ursula K. Le Guin** (1929–2018)
Although known primarily as a writer of speculative fiction, Le Guin's literary worlds provided profound commentary on cultural complexity, ecological balance, and the fluidity of identity. By imagining societies with radically different norms and ecosystems with unusual dynamics, she compelled readers to reconsider what is "natural" or "inevitable." Through metaphor and narrative, Le Guin showed that complexity isn't confined to biology or physics—it also thrives in the realm of ideas, stories, and human choices. Her work encourages us to cherish diversity, question assumptions, and appreciate how meaning arises from shifting perspectives.

4. **Ilya Prigogine** (1917–2003)
A chemist and physicist celebrated for his work on non-equilibrium thermodynamics, Prigogine explored how order can arise far from equilibrium. His insights into dissipative structures—ordered patterns emerging in turbulent systems—demonstrated that entropy's increase doesn't always spell a slide into chaos. Instead, conditions can spawn new forms of organization, mirroring life's ability to harness energy and evolve complexity. Prigogine's theories bridged chemistry, physics, and philosophy, challenging the notion that systems inevitably degrade into disorder and reinforcing the idea that innovation can flourish under unstable conditions.

5. **Jane Goodall** (1934–)
A primatologist and conservationist, Jane Goodall redefined the relationship between humans and other primates. Her long-term field studies in Gombe Stream National Park revealed that chimpanzees use tools, form social bonds, and demonstrate cultural behaviors once thought uniquely human. By immersing herself in the community of another species, Goodall challenged the boundaries that separate humans from the rest of life on Earth. Her work highlights how complexity arises not only from physical processes and ecological interactions but also from the rich inner lives and relationships of living beings. Goodall's patient, empathetic approach

encourages us to appreciate nuance, continuity, and individuality in the tapestry of life.

6. Stuart Kauffman (1939–)

A physician turned theoretical biologist, Kauffman delved into the origins of life and the principles of self-organization. Through models of "autocatalytic sets" and "rugged fitness landscapes," he showed how spontaneous order can emerge from chaos, generating life's richness without external design. His interdisciplinary mindset—drawing on biology, mathematics, physics, and philosophy—offered new ways to think about complexity, emphasizing that the biosphere is not a static entity but a ceaseless creative process. Kauffman's vision inspires us to see life as an ongoing discovery, an evolving experiment where novelty is always possible.

7. Buckminster Fuller (1895–1983)

An architect, designer, and inventor, Fuller approached technology and resource use as global puzzles requiring holistic solutions. His geodesic domes, Dymaxion maps, and forward-thinking proposals reflected a desire to harmonize human structures with nature's patterns of efficiency. Although not all of his ideas took root, his willingness to blend engineering, ecology, and social responsibility embodied a form of creative complexity. Fuller's legacy reminds us that the quest for sustainable, integrated systems is itself an act of imagination—an experiment in shaping complexity that respects both human aspirations and natural constraints.

8. **Maria Sibylla Merian** (1647–1717)

Long before ecology became a formal science, Merian's meticulous illustrations of insects and plants revealed complex life cycles and ecological relationships. In an era dominated by static taxonomies, she depicted transformations—caterpillars becoming butterflies, eggs becoming larvae—as dynamic, interconnected processes. Her work bridged art and science, offering a visual narrative of complexity in action. Merian's approach demonstrates that careful observation and creative depiction can uncover patterns otherwise overlooked, blending empirical rigor with aesthetic sensitivity.

9. **Wangari Maathai** (1940–2011)

A Kenyan environmental activist and founder of the Green Belt Movement, Maathai recognized the interplay between ecological health, social justice, and cultural vitality. By mobilizing communities to plant trees, she promoted biodiversity, soil stability, and economic resilience. Her work showed that healing damaged landscapes involves more than just restoring vegetation—it requires understanding how human values, traditions, and livelihoods intertwine with the natural world. Maathai's legacy attests to the notion that complexity includes human actions and moral choices;

we shape and are shaped by our surroundings in countless ways.

10. **Murray Gell-Mann** (1929–2019)

A theoretical physicist who identified the "quark" as a fundamental building block of matter, Gell-Mann also took a wide-ranging interest in complexity. Through involvement with the Santa Fe Institute and collaborations across disciplines, he explored how simple rules give rise to elaborate patterns in nature, language, and culture. Gell-Mann's curiosity and intellectual rigor exemplify the rewards of venturing beyond one's specialty. He reminds us that complexity is a universal theme, linking the quantum world to ecological networks and human societies, and that understanding it requires both depth and breadth.

Collectively, these figures underscore that complexity is not just a topic of study, but a lived experience—one enriched by perseverance, imagination, empathy, and courage. Their contributions resonate with the themes woven through this book: that life's creativity arises amid tension and uncertainty, and that to engage with complexity is to embrace a universe in perpetual motion. Each thinker, in their own way, has shown us that no matter the domain—science, art, philosophy, activism—the invitation to explore and co-create complexity remains open.

Conclusion of the Appendices

As we close these supplementary sections, it's worth reflecting on their purpose. The tools, references, and figures gathered here are not endpoints, but gateways—starting points for deeper inquiry. They underscore that complexity, in all its many forms, cannot be fully contained within the pages of any single book. Instead, it requires a tapestry of insights: conceptual frameworks that clarify the subtle interplay between order and chaos, scientific data that ground our understanding in tangible evidence, philosophical reflections that probe the meaning of it all, and creative, interdisciplinary leaps that allow us to connect seemingly disparate dots.

The figures we've highlighted remind us that complexity emerges not only from matter and energy, but also from the courage and curiosity of the human mind. Each thinker, artist, or scientist forged new paths, often without a clear roadmap, illuminating truths that might have remained hidden. Their legacies stand as invitations for us to engage, adapt, question, and imagine anew.

As you carry these insights forward—into your own research, dialogues, creative endeavors, or contemplative moments—remember that no collection of texts or profiles can capture the full richness of an evolving cosmos. Rather than seeking ultimate answers, we can strive for richer questions; instead of clinging to fixed models, we can remain open to revision and surprise. In this spirit, these appendices and the book as a whole encourage an ongoing conversation—one that draws on our collective intelligence and imagination as we navigate a universe that never ceases to unfold fresh patterns.

May these resources serve as stepping stones, and may the path you take from here lead to discoveries not yet imagined. Complexity lives in the relationships we cultivate, the perspectives we explore, and the humility we bring to every encounter with the world's shifting mosaic of possibilities.

Conclusion of the Book

Step outside, and let your senses wander. Whether you find yourself in a bustling city or a quiet corner of nature, the world is alive with subtle conversations between chaos and order. The interactions you witness—sunlight filtering through leaves, people weaving through crowded streets, ideas passing through minds—are all part of an ongoing interplay that stretches back through cosmic time. Nothing stands still. Every pattern that emerges stands poised at the edge of transformation, every form we know remains open-ended, ready to shift in ways we cannot fully predict.

In these pages, we explored how life, from its earliest sparks, has tapped into this dynamism to coax complexity from a universe inclined toward disorder. We traced the journey through entropy's relentless push, the quantum intricacies of green leaves, the creativity of evolutionary processes, and the ways human ingenuity joins the dance. Along the way, we discovered that structure does not arise in defiance of nature's laws, but in conversation with them. Complexity emerges where forces meet and adapt, where environments change and living systems respond, where minds grapple with questions that have no final answers.

This perspective frees us from the notion that the cosmos should provide neat endings or perfect solutions. Instead, it offers a ceaseless invitation to engage. Complexity is not a prize we win or a puzzle we solve once and for all;

it's an evolving phenomenon that thrives on new inputs, fresh insights, and the interplay between multiple scales and contexts. The universe whispers across star fields and forest floors, between chemical bonds and cultural exchanges, suggesting that meaning can be found not in static certainty, but in ongoing participation.

As readers, thinkers, and co-creators, we add our voices to this chorus. The capacity to perceive patterns, ask questions, and imagine alternatives belongs to all of us. Each generation inherits a legacy of knowledge, but also a horizon of unknowns, and each has the opportunity to push that horizon a bit further. In acknowledging that complexity is forever in flux, we grant ourselves permission to remain curious, to embrace uncertainty, and to find beauty in what we do not yet comprehend.

We need not fear that meaning will slip away without a final verdict. Instead, we can find comfort and inspiration in the fact that life, at every turn, has proven capable of forging new forms of order in the face of dissolution. From a cosmic perspective, our moments of insight, creation, and care are part of a grand tapestry that outlives any single narrative. The stories we tell—through science, myth, art, and philosophy—reflect our best attempts to trace the contours of an ever-changing reality.

As we move beyond this conclusion, the patterns we've observed and the voices we've highlighted do not vanish; they accompany us as we navigate our lives, communities, and endeavors. They remind us that every plan, insight, and invention we cherish will be shaped by forces beyond our control. Yet, this inevitability is not a curse—it's an invitation to constant learning and reinvention. Complexity is not fragile; it is resilient and fertile, thriving on challenge and transformation.

Rather than closing the book on these themes, we can consider this the start of a more subtle and enduring engagement. You may find echoes of these ideas in a conversation at twilight, in the rustle of leaves outside your window, or in a spark of understanding that arises while pursuing your craft or profession. Complexity meets us everywhere, and we are free to respond—to sculpt our understanding, refine our questions, and weave new threads of meaning.

In acknowledging that no single framework or narrative can capture the full sweep of existence, we set ourselves free from the search for ultimate conclusions. Instead, we become collaborators in an ongoing project—one that life itself has been undertaking since its earliest beginnings. Each act of comprehension, each insight, and each moment of empathy adds another layer of possibility to the tapestry. The universe invites us to dance with it, to play at the edges of known and unknown, and to welcome emergence as our constant companion.

There is no prescriptive map charting a path forward, only the reassurance that we are capable of engaging thoughtfully, compassionately, and creatively with complexity. The cosmos did not deliver a set of instructions; it entrusted us with the ability to sense patterns, adapt, and find meaning in our own shared endeavors. In that shared endeavor, we are all sculptors, shaping and reshaping the contours of our understanding, contributing to a complexity that envelops and transcends our individual lives.

Wherever you go from here—whatever fields you study, communities you join, or challenges you embrace—these insights can serve as subtle guides, gentle reminders that the world around us is forever in motion. Though we cannot predict exactly where complexity will lead, we can trust that new patterns will arise. And in that trust lies a quiet confidence: that we, as participants in this vast interplay, can continue the dialogue, deepening our appreciation for a universe that never stops whispering new possibilities.

This book may reach its final page, but the exploration never truly ends. May this conclusion serve not as a closing door, but as a threshold through which you pass, carrying forward the spirit of inquiry and wonder into every unfolding chapter of existence that lies ahead.

So, close this book not to shelve it indefinitely, but to carry its spirit into your days. Let the notion that complexity emerges and evolves stay with you, a subtle lens altering how you notice patterns in your work, your relationships, and the natural world around you. Let curiosity remain a guiding star, unafraid to venture into realms where clarity fades into mystery. Let humility temper your convictions, for all knowledge is provisional and open to revision. And let creativity spark within you, prompting you to shape new forms, tell new stories, and embrace the fluid interplay of order and chaos that defines our existence.

In the grand scheme of things, we are all sculptors—albeit transient ones—contributing to an ongoing masterpiece that no single mind can envision in its entirety. Life, in its myriad expressions, coaxes complexity from a universe ever on the cusp of change. By observing, questioning, and engaging, we become part of that coaxing, part of the whisper's echoing chorus. It is an honor and a responsibility, one that binds us to each other and to the cosmos itself.

Now, step forward into the unknown. The dance between structure and unstructured forces awaits your participation. Complexity beckons like a distant melody, inviting you to listen, respond, and create anew. There are no final notes, only the promise of more patterns to discover, more meaning to craft, and more futures to imagine.

Hey!! The time is?? NOW!!

To enhance your reading experience, I've added a playlist of related videos on YouTube that further explore the concepts in this book—featuring expert interviews, guided exercises, and more. Feel free to scan the QR codes to watch them and deepen your understanding.

Explore Further on YouTube

As these topics continue to evolve, I'll update the book—So please **follow me on Amazon** to stay tuned.

Last thing, Please feel free to share your thoughts with me via my **Telegram**. Use this QR code to connect with me:

Thank you for joining me. I hope it sparks your curiosity and inspires you to keep exploring!

www.ingramcontent.com/pod-product-compliance
Lightning Source LLC
Chambersburg PA
CBHW050317230526
45471CB00005B/2218